METHODS FOR PREPARATION AND SCREENING OF *r*-DNA

METHODS FOR PREPARATION AND SCREENING OF *r*-DNA

By

Dr. Narendra Srivastava
Ph.D, FZSEI, CMLT, DCA
Univ. Dept. of Zoology
B.N. Mandal University
Madhepura (Bihar)

Dr. Arun Kumar
Ph.D
Univ. Dept. of Zoology
B.N. Mandal University
Madhepura (Bihar)

UNIVERSITY SCIENCE PRESS
(An Imprint of Laxmi Publications Pvt. Ltd.)
An ISO 9001:2015 Company
BENGALURU • CHENNAI • GUWAHATI • HYDERABAD • JALANDHAR
KOCHI • KOLKATA • LUCKNOW • MUMBAI • RANCHI
NEW DELHI

Methods For Preparation and Screening of r-DNA

Printed and bound in India
Typeset at: Sukuvisa, New Delhi.
First Edition : 2011
ISBN 978-93-80856-61-2

Branches

©	Bengaluru	080-26 75 69 30	
©	Chennai	044-24 34 47 26,	24 35 95 07
©	Guwahati	0361-254 36 69,	251 38 81
©	Hyderabad	040-27 55 53 83,	27 55 53 93
©	Jalandhar	0181-222 12 72	
©	Kochi	0484-237 70 04,	405 13 03
©	Kolkata	033-22 27 43 84	
©	Lucknow	0522-220 99 16	
©	Mumbai	022-24 93 12 61,	
©	Ranchi	0651-220 44 64	

Published in India by

UNIVERSITY SCIENCE PRESS
(An Imprint of Laxmi Publications Pvt.Ltd.)

An ISO 9001:2015 Company
113, GOLDEN HOUSE, DARYAGANJ,
NEW DELHI - 110002, INDIA
Telephone : 91-11-4353 2500, 4353 2501
Fax : 91-11-2325 2572, 4353 2528
www.laxmipublications.com info@laxmipublications.com

C—2534/010/11
Printed at: Ajit Printers, Delhi.

CONTENTS

B.N.Mandal University

Madhepura-852113 (Bihar)

Prof.R.P. Srivastava
Ph.D.,D.litt.
Vice-Chancellor

06476-222059
06476-222578 (Fax)
Mob.:-09431810309
Email : vc_bnmu@yahoo.com

Ref..........................

Date..........................

FOREWORD

We are passing through an era of post-postmodernism which encompasses various elements of technological developments. The most important of these developments is bio-technology which is a grand combination of Biology and Technology. Technology, today, has made its dent into all walks of life. Its inclusion in the realm of Biology has given birth to numerous branches of scientific knowledge which are highly beneficial for human civilization and culture. In this period of globalization the intermixing of human culture is only a logical corollary of the existential behavior of man. The academicians who propagate the theories of evolution, particularly environmental evolutionism, are specially proud of having the outcome of researches of bio-technology.

The present book is more or less a glossary of the details of this latest addition of scientific achievement, that is, the bio-technology. The author of the book happens to be a member of the family of the university I am presently heading. I am hardly acquainted with the scientific developments encircling the global existence and naturally have no entitlement for passing any comment on the merit of the book. But since I am personally acquainted, rather impressed by the academic excellence of the author I do hope, perhaps rightly, that the book will prove to be extremely helpful for those who want to acquire much needed knowledge of Bio-technology.

I wish all my good wishes for the author Dr. Narendra Srivastava. May his academic interest shine more prominently in time to come.

(Ripusudan Pd. Srivastava)

Dr. T.V.R.K. RAO
Ph.D; F.I.C; LL.B
Chartered Chemist
Reader in Chemistry
ASSOCIATE EDITOR (BIOCHEMISTRY SECTION)
INDIAN CHEMICAL SOCIETY, KOLKATA
e-mail: tvrk_rao@rediffmail.com

PREVIEW

The book entitled **The Method for Preparation and Screening of r-DNA** is an endeavor for a lucid presentation of the subject is an easily assimilable form. Various methodologies concerning the preparation, screening and recovery involved in recombinant DNA Technology have been discussed in optimal details with a good clarity. Distribution of the work into various chapters is quite systematic and sequential. Figures and drawings are well labelled and illustrative. Various Appendices placed in the book are indeed useful for the practiceners of the r-DNA Technology. From bibliographic angle, the book depicts a thorough coverage with the inclusion of some of the very recent references. Author of the book, Dr. Narendra Srivastva has, seemingly put up hard labor and spent much time at his study table in bringing out this volume. Author's own experiences in Teaching and Reasearch in the Post Graduate classes are also latently reflected in the book. Hopefully, the book would prove to be a good academic menu for the teachers and researchers dealing with the subject of "Recombinant DNA Technology". I wish the book a successful journey into the hearts of its readers.

(T.V.R.K. Rao)

PREFACE

The present century is recognized as an era of biotechnology as it provides all basic scientific tools responsible for revolutionalization of global civilization and economy. It is now established as a discipline and gradually developing as a research oriented science. It is basically a combination of Biology and Technology and covers subjects like genetics, biochemistry, microbiology, cytology, biometry, medicine, agriculture, animal husbandry etc. All biotechnological works are fundamentally determined by the changing of naturally available genome of living system, known as genetic engineering or technology of recombinant DNA preparation. At present most of the universities or institutes do not offer courses in genetic engineering as a separate discipline but as a subsidiary to biotechnology, microbiology and biochemistry streams. In India, Biotechnology began during mid 1980's and was concentrated within the large domestic and multinational drug firms. But now Government of India and its associated academic or scientific bodies like UGC, CSIR, DBT *etc.*, have launched a number of programs and proposals to accelerate the output of expertise personals in this field.

In spite of all efforts on behalf of government and its academic and scientific bodies, the growth rate in the field of biotechnology and genetic engineering is not in perfect pace with global development due to lack of basic infrastructure in the laboratories of colleges or universities and knowledge of fundamental techniques for genetic engineering. The present book entitled "Methods for Preparation and Screening of r-DNA" is prepared in keeping with this view. The present book describes basic and simple methods for extraction of nucleic acids, preparation of recombinant DNA and screening of clones in moderate laboratories of colleges and universities. Each experimental method is preceded by a brief introduction, describing fundamental theoretical concept necessary for laboratory work. The language of the book is simple and lucid and steps pertaining to each laboratory method are easily understandable. The book is useful for common students of life sciences, agriculture, veterinary science and medical sciences who want to initiate their career in the field of Biotechnology/Genetic Engineering.

The book is prepared with full attention however author cannot overrule the possibility of printing errors. The author will cordially accept all constructive suggestions necessary for improvement of the quality of the book and its usefulness.

—**Authors**

ACKNOWLEDGEMENT

We are thankful to all those teachers and laboratory staff who extended their valuable help in proper understanding of all experimental methods depicted in the present book. I am particularly thankful to Dr. Anita Sironmani, Dept. of Biotechnology, Kamraj Madurai University, Tamil Nadu and Dr. Ashok Gupta, Dept. of Biochemistry, Patna University, Bihar for academic discussion and help in preparation of book. We are thankful to our student Mr Ravindra Shrivastava, Lecturer, Insan College, Kishanganj, Bihar for his help in preparation of manuscript. We pay my special respect to Vice-Chancellor Dr. Ripusudan Prasad Srivastava for their affection and encouragement for academic pursuits. We are also thankful to Vice-Chancellor, M.M.H.A.P University, Patna, Dr. Quamar Ahsan; Dr M.P. Sinha, Dept of Zoology, Ranchi University, Dr. T.V.R.K Rao, Associate Editor (Biochemistry Section), Indian Chemical Society, Kolkata and Dr. (Mrs) Latika Sharan, Dept of Botany, Ranchi Womens College, Ranchi, Jharkhand. I have immense regard for my brothers Sri Ashok, Sri Dileep Kumar Prasad and Rajnikant Prasad and deep affinity and love for my wife Mrs Babita Sinha, who all along kept on encouraging me towards actualizing the academic pursuit. We are also thankful to Mr. Shambhu Narayan Yadav, PA cum Secretary to VC and Md Naziruddin, DSW, B.N.Mandal University for their brotherhood affection and encouragement.

—Authors

Chapter 1

RECOMBINANT DNA : AT A GLANCE

1.1 INTRODUCTION

Genetic engineering is an advanced branch of science. It involves a wide range of techniques used for changing of nucleotidic sequence in the DNA code in a living organism. "Genetic engineering" refers to deliberate modification of the characters of an organism by the manipulation of its genetic material. It comes under the broad heading of biotechnology. The term biotechnology seems to have been first used by the Leeds City Council in the United Kingdom in the early 1920's when it established an "Institute of Biotechnology". In 1972, *Berg et al* combined DNA from two viruses to produce what is called recombinant DNA. It was *Chakravarty* (1973) who isolated genes from one organism and inserted them into another for the purpose of degrading oil and provided the basic technique for genetic engineering. There is a great scope in this field as the demand for genetic engineers are growing in India as well as abroad.

The three basic components of genetic engineering are desired gene, vector and cloning cell. The desired gene is obtained either from natural living components or from extracted *m*-RNA by the method of reverse transcription or from vitro synthesis in laboratory. At first, *Dr. Har Govind Khurana* synthesized artificial gene in laboratory corresponding to Ala-*t*RNA. The common method for collection of desired gene is through extraction of genomic DNA.

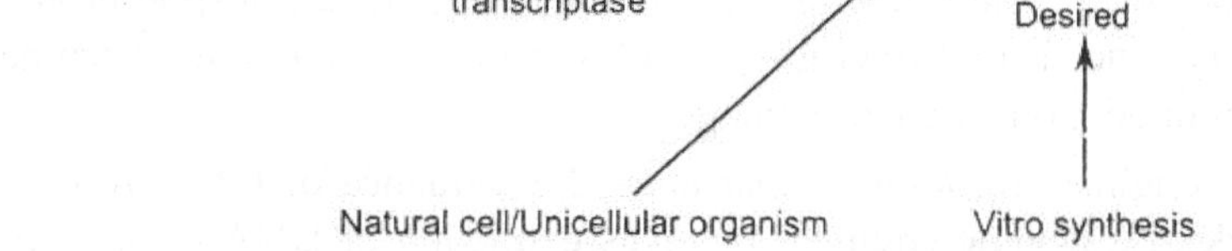

FIGURE 1.1. *Represents sources for desired gene*

The details of method are described in later chapter of this book. For the collection of desired gene by reverse transcriptase at first *m*-RNA is extracted by subcellular fraction. The method used for subcellular fractionation is differential centrifugation. The purified *m*-RNA is used as a template to synthesize complementary DNA. The *c*-DNA is single stranded and the process is mediated by enzyme reverse transcriptase. The enzyme reverse transcriptase was at first reported by *Baltimore* and *Temine* in retrovirus. The synthesised hybrid of *ss*-DNA —*m*RNA is separated by alkali hydrolysis. The separated *ss*-DNA is treated with DNA polymerase to synthesise *ds*-DNA but with one continuous end due to formation of regular 5′- 3′ phosphodiester bond during the process of polymerization. The continuous end is separated by enzyme s_1 nuclease and both strands are free but remain associated by complementary nucleotidic hydrogen bonds. The ends of synthesized DNA have either sticky ends or blunt ends. In later condition, adenylic polynucleotid or thymidylic polynucleotid sequence is added to 3′ ends of both strands for the making of sticky ends. The process is known as homopolymer tailing. It is endergonic process and requires ATP or TTP as a source of energy. The process is mediated by enzyme terminal transferase.

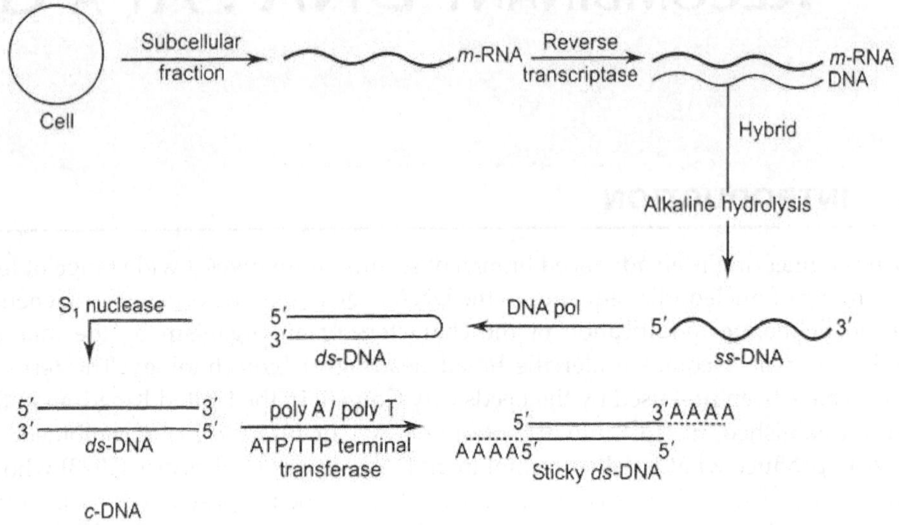

FIGURE 1.2. *Steps for synthesis of DNA by Reverse transcriptase*

The desired gene or fragment of genomic DNA is then inserted into genetic material of other living organisms for propagation and survival. The genetic material, which we use for this purpose, is known as vector. The vector used commonly in laboratory is plasmid, has some specific characteristics. It is collected from bacteria by ultracentrifugation. The plasmids are treated with enzyme restriction endonuclease. The action of enzyme is sequence specific and cuts 4–6 nucleotidic sequence known as palandrom. The blunt end of plasmid is also converted into sticky by homopolymer tailing. Finally we add desired gene or fragment of DNA to vector used for cloning by the action of enzyme DNA ligase. The enzyme forms a phosphodiester linkage between free 5′ phosphoryl and 3′ hydroxyl group and join both components. Commercially enzyme DNA ligase is obtained from λ bacteriophage.

Genetic engineering aims to rearrange the sequence of DNA by use of artificial methods. Hence work of a genetic engineer involves extraction of DNA from an organism, changing its nature by the use of chemicals or radiation and subsequently putting it back into the same or a different organism. For example, genes and segments of DNA from one species is taken and put into another species. They also study how traits and characteristics are transmitted through generations and how genetic disorders are cured. Genetic engineering in plants and animals may

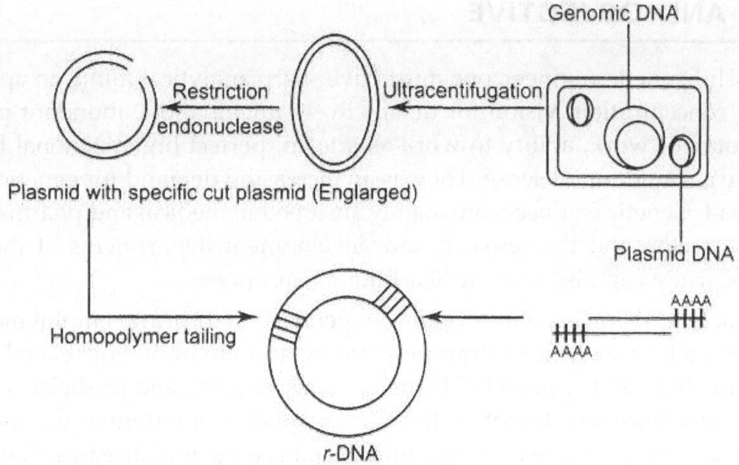

FIGURE 1.3. *Steps for extraction and preparation of recombinant DNA*

be used to improve certain natural characteristics or value to increase resistance to disease or damage and to develop new characteristic. It is used to change color, size, texture etc., of plants otherwise known as GM (genetically modified) foods. Genetic engineering in human can be used to correct severe hereditary defects by introducing normal genes into cells in place of missing or defective ones. By the use of genetic engineering even we are developed transgenic animals also having desirous features, known as transgenic animals.

A team headed by *Ivan Wilmut* and his colleagues at the Roslin Institute in Edinburgh, Scotland made history when they produced a lamb named Dolly, an exact genetic copy or clone of a sheep. This landmark discovery of the regeneration of an exact replica of a whole animal by transferring nuclei from the cells of that animal, without the help of a male counterpart, has given researches a wide areas open to be discovered. With this discovery, genetic engineering has become globally recognized.

Genetic engineering comes under the broad area of biotechnology. The biotechnology industry depends on genetics to produce pharmaceutical products such as insulin, somatotrophin, *t*-PA, lymphokines, Interferon, and other therapeutic chemicals. In the medical field, genetics focuses on genetic diseases. It strives to understand the molecular basis of diseases and their cure. The present thought is responsible for the initiation and development of gene therapy. Genetic tests help in identifying culprits in criminal cases. Due to the more and more applicability of genetic engineering at socio-economic level of human being, the branch has a fast growing area and has thrown up several job opportunities. At present most of the universities and institutes do not offer courses in genetic engineering as a separate discipline, but as a subsidiary to biotechnology, microbiology and biochemistry streams. Undergraduate and postgraduate courses in Biotechnology offer specialization in genetic engineering. In spite of a number of advantages of genetic engineering, it has some negative impacts also. The most significant is the chance of loss of genetic diversity and existence of a condition leads to loss of a gene from gene pool due to selection of a particular gene. Nowaday scientists are discussing on the ethical problems, can be caused due to more and more popularizing of this branch. A balance between naturally occurring living components and genetically altered living components is necessary for the regularization and maintenance of living world with existing genetic diversity.

1.2 AIM AND OBJECTIVE

To be a successful genetic engineer, one must have sharp analytical mind, an aptitude for research, high levels of concentration, vision for details lively imagination, abundant physical stamina to put in long hours of work, ability to work as a team, perfect organizational behavior moreover he should have a sound moral sense. There is an increasing demand for genetic engineers in India as well as abroad. Genetic engineers are mainly absorbed in medical and pharmaceutical industries, the agriculture sector, and the research and development departments of the government and private sectors. They can also take up teaching as an option.

Besides teaching the main opportunities in genetic engineering, employment avenues in this field lies in research in biotech laboratories, energy and environment related industries, animal husbandry, agriculture and applied fields such as diary farming and medicine. Genetic engineering is that wonderful science which enables the human minds to interfere in and modify the processes of life, birth, death and even offers escape from certain congenital diseases. Some of the important institutes of India, actively engaged in teaching and research are mentioned below :

(*i*) The National Institute of Immunology (NII), New Delhi.
(*ii*) The Centre for DNA Fingerprint and Diagnostic, Hyderabad.
(*iii*) Biochemical Engineering Research and Process Development Centre, Chandigarh.
(*iv*) The Institute of Genomics and Integrative Biology, Delhi.
(*v*) All India Institute of Medical Sciences, New Delhi.
(*vi*) Molecular and Human Genetics, Banaras Hindu University, Varanasi.
(*vii*) Microbial Gene Technology, Madurai Kamraj University, Madurai.
(*viii*) The School of Basic Medical Sciences, University of Madras, Chennai.
(*ix*) The Indian Institute of Technology (IIT), Delhi, Guwahati, Kharagpur and Madras.
(*x*) Centre for Biotechnology, JNU, New Delhi.
(*xi*) Human Genetics, Gurunanak Dev University, Amritsar.
(*xii*) Rajendra Agriculture University, Pusa, Samastipur, Bihar.
(*xiii*) Department of Biotechnology, Himanchal Pradesh University, Shimla.
(*xiv*) Chawdhary Charan Singh Haryana University, Hissar.
(*xv*) Biomedical Genetics, Vellore Institute of Technology, Vellore.

1.3 SAFETY MEASURES

The genetic engineering is a fast growing area of modern education, attracts a large number of youngsters. The study area is lab intensive and fundamentally associated with microbial populations and hazardous chemicals having no or poor knowledge of toxicological properties. Hence it is necessary for students and workers to maintain proper safety measures in laboratory. Some important points are mentioned here for guideline.

1. **Laboratory Sterilization :** The sterilization of laboratory both before and after of laboratory work is necessary. The students and workers are advised to use separate aprons and sleepers for laboratory. The lab bench and lab walls must be free from holes and gaps. The doors and windows are sealed to prevent contamination of both direction, that is from outside to laboratory and vice versa.

The department of health, education and welfare has classified various bacteria into different categories with regard to shipping requirements. Nonpathogenic strain of (K12) and *B.subtilis* are in class-I and are considered to present no or minimal hazard under normal shipping conditions. However *Salmonalla, Haemophilus* and certain strains of Streptomyces and *Pseudomonous* are in class-II. Class-II bacteria are agents of varying degree of severity but are contained by ordinary laboratory techniques.

2. **Use of Sterilizing Instruments :** Autoclaves and microwave ovens in the lab require certain precautions. Accidents have occurred involving their use (*e.g.*, melting agar or bactro-agar stored in bottle or to sterilize). If the screw top is not completely removed and there is not enough space for the steam to vent, the bottle can explode and cause severe injury when the containers are removed from the microwave or autoclave. Always completely remove bottle caps before microwaving or autoclaving. An alternative method for routine agarose gels that do not require sterile agar is to weigh out the agar and place the solution in flask.
3. **Use of Bleach :** The lab bench is always clean by bleach after work in laboratory. It is used as disinfectant. But it is poisonous, can be explosive and may react with organic solvent. It may be fatal for inhalation and is also harmful by ingestion and destructive to skin. Wear appropriate gloves and safety glasses and use in a chemical fume level to minimize exposure and odor.
4. **Careful Use of Electrical Instruments :** The instruments associated with power supply, like electrophoretic instrument and spectrophotometer, pose serious fire hazards and electric shock hazards if not used properly.
5. **Careful Use of Chemicals :** The chemicals used in laboratory works are sometimes harmful like glycine, coomassie brilliant blue, methylene blue, phenol red, boric acid, magnesium chloride, isopropanol, bromophenol blue, Tris, acetic acid, ammonium sulphate etc. Must use gloves during handling of such type of chemicals. The enzyme lysozyme is coarsive to mucous membranes.
6. **Protect from Chemical and Physical Mutagenic Agent :** During experimental works, there is chance of exposure to mutagenic chemical and physical factors like ethidium bromide and UV rays. Ethidium bromide is a powerful mutagen and is toxic. Avoid breathing the dust. The source of UV light is transilluminator. It is dangerous and can damage the retina of the eyes. Always use filter or safety glass when observe nucleic acids in transilluminator.

Chapter 2

EXTRACTION OF DNA

2.1 INTRODUCTION

Watson and Crick presented the structural model of DNA in 1953–56. But their achievement was based on the work of a number of contemporary workers like *Linus Pauling, Chargaff et al, William Astbury, Maurice Wilkines, Rosalind Franklin* etc. The presentation of *Watson* and *Crick* has been revolutionized the branch of molecular biology and nowaday works on nucleic acids are commonly conducted in laboratories of graduate and postgraduate courses. Many techniques are currently available that allow the isolation of DNA or RNA. Most of the procedure is taken the advantage of the fact that phenol can act as an efficient deproteinizing agent quickly disrupting cellular integrity and denaturing proteins. The work of genetic engineering is initiated by the extraction of nucleic acids.

The method utilizes an initial phenol extraction coupled with phenol: chloroform extraction to simultaneously remove proteins and lipids from nucleic acid containing solutions. Also the composition of the aqueous extraction buffer is optimized to increases nucleic acid recovery. The integrity of the DNA/RNA will be improved by maintaining harvested cells or tissues in cold.

2.2 CHEMISTRY OF DNA

DNA is a complex organic compound and its properties are determined by both chemical nature of structural components and configuration. Before the presentation of model by *Watson* and *Crick, Alexander Toad* (1951) *et al* was at first suggested that nucleiotides are combined by phosphodiester bond, one phosphate bond is formed with 5'-OH carbon of sugar and other phosphate bond is formed with 3'-OH carbon of sugar. DNA is configurationally flexible and its various forms have been known to science. Even X-ray crystallograghy pictures of Rosalind Franklin represented its two (A and B) forms. The B-DNA configuration has been described by *Watson* and *Crick* characterized by right handed interwoven double helix with alternatively major and minor grooves.

The primary structure of DNA is formed by polymerization of 2' - deoxyribo nucleotides. It contains an aromatic nucleobase. The polymerizing bond is 5' → 3' phosphodiester bond. It is formed by the hydrolysis between C3'-OH of pentose sugar and one of the OH of phosphoric acid. The phosphoric acid is responsible for negative charge of DNA backbone. The high density of negative charge associated with this biopolymer also dictates to some extent the character of agents which associated with it, agents that bind most efficiently to DNA often possess a net overall positive charge. In eukaryotic cell, the binding of histone protein is best and common example for such type of binding. The polyanionic backbone provides a chemical condition for the formation of non-covalent interactions significant for structural stability. The anionic phosphates of the backbone interact electrostatistically with cations Na^+/Mg^{2+} and reduce phosphate - phosphate repulsive interaction and eventually increase the configurational stability of DNA.

The structural configuration of primary DNA strand, has negatively charged backbone and hydrophobic nucleobase, is relatively stable. In DNA, postranscriptional chemical modification is rarely observed. For example, the DNAs of eukaryotes contain relatively simple modifications involving the methylation of the 5th position of cytosine or the exocyclic amino group of adenine. The modifications have gene regulatory effect. The methylated segment of DNA is inactivated and non-functional. It is non-responsive to enzymatic activity.

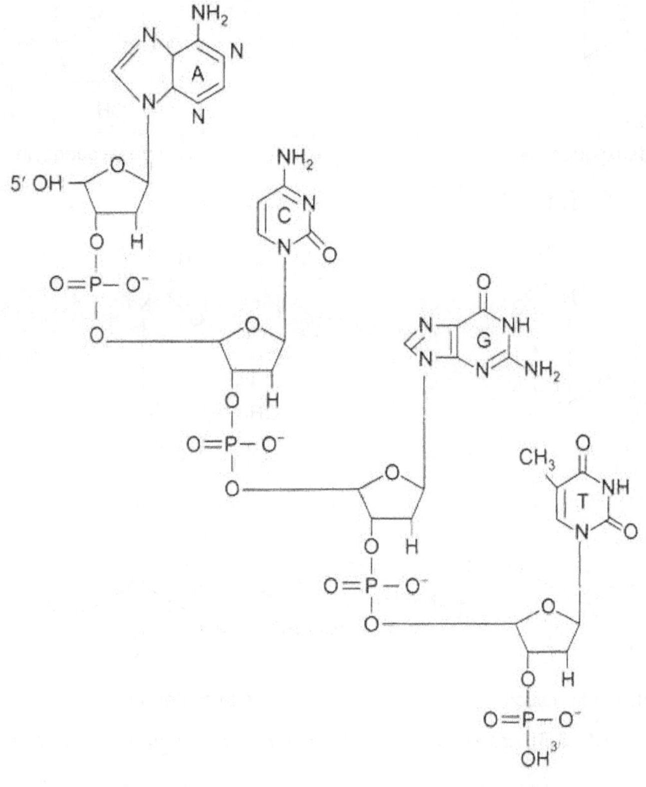

FIGURE 2.1. *The primary structure of DNA*

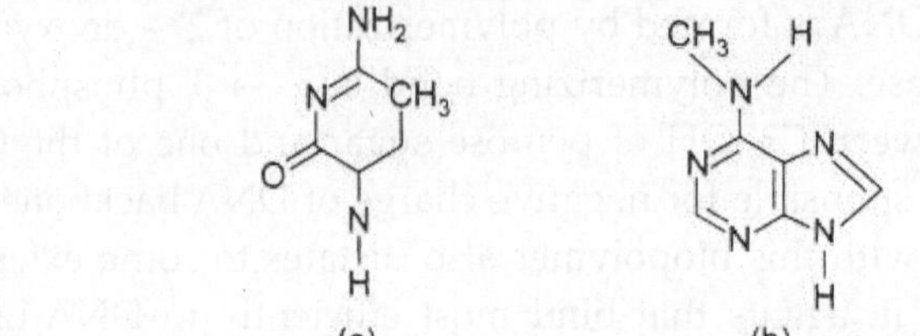

FIGURE 2.2. *Represents 5-methyldeoxycytidine (a) (m^5dc) and 6-methyldeoxyadenosin (b) (m^6da)*

The other chemical specifications which determine configuration of DNA are deoxyribose ring configuration and the N-glycosidic bond angles. In active configuration of DNA the deoxyribose ring adopts a C2'-endo conformation, while the N-glycosidic bond angle is in an anti-configuration.

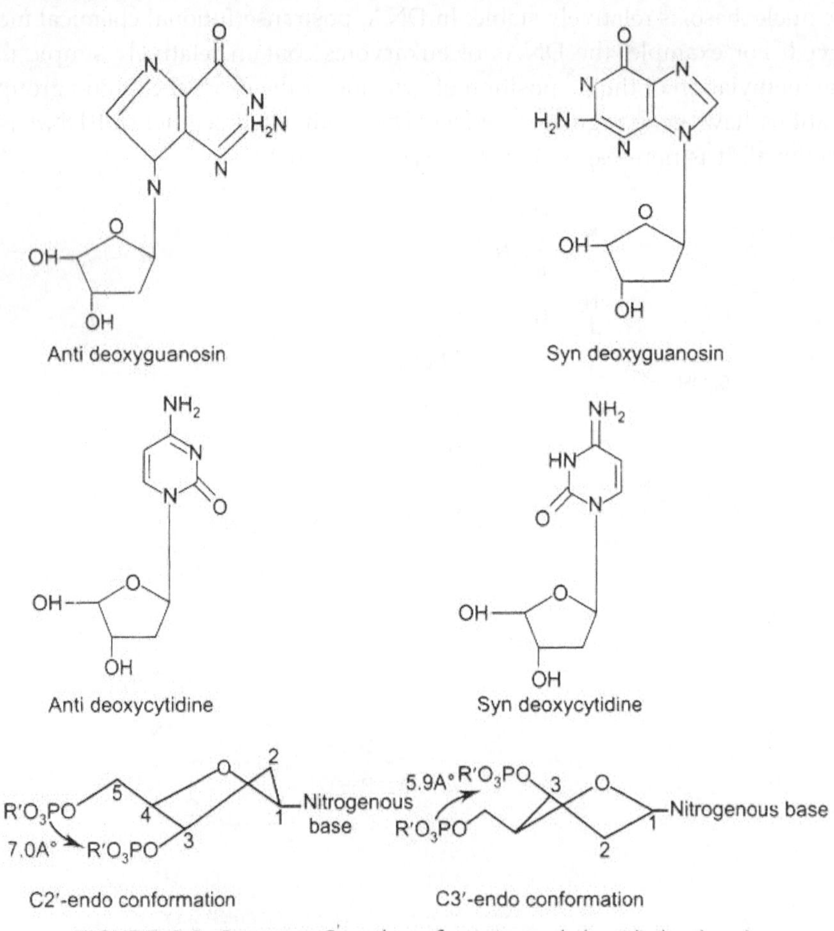

FIGURE 2.3. *Represents C_2'-endo conformation and glycosidic bond angle*

The structurally observable minor groove and major groove have also chemical significance as these two grooves create unique microenvironments for the binding and recognition of ligands. The floors of the grooves of the DNA helix would differ for individual base pair sequences with respect to their patterns of H-bond donors, H-bond acceptors, and sites available for hydrophobic interactions due to presence of C5 methyl group of thymidine. This microheterogeneity of the DNA results from localized, sequence dependent changes in the shape of the helix which acts as limiting factor for steric interactions between the purines of adjacent base pairs. In active form of DNA, the minor groove is the preferred binding site of low molecular weight species such as the antitumour agents distamycin and netropsin. The size and chemical features of these particular drugs complement the narrow minor groove of AT rich DNA regions results a tight and sequence selective binding interaction.

The chemical nature of DNA described is based on B form although it is polymorphic. The recent research publications also reported some unusual structures of DNA. These structures are often highly localized to a particular base sequences and require a precise set of conditions to extrude from B form. Some examples are cruciforms observe within palindromic DNA sequences (*Lilley*, 1989), triple stranded structure (*Rajegopal and Feigon*, 1989), H form structures (*Htun* and *Dallberg*, 1988), and quartet structures consisting of four DNA strands (*Guschlbaner et. al.* 1990). One of the unusual DNA is observed in structure of telomeric DNA which plays an important role in stabilizing chromosomes. The telomeres refer to the terminal protein DNA complexes of the chromosomes, and the DNA sequences found in telomeres are rich in G and contain a short, highly repetitive sequences such as (TTGGGG)n. The study after crystallization of telomeric DNA (*Kang et al* 1992) represents hairpin like structure, and two of these hairpin joined to make a four stranded helical structure with the thymines forming loops at either end. The four structure formed by the guanine residues being held together by a cyclic hydrogen bonding network and a metal ion (K^+) located at the centre. The four guanine residues are found to have a glycosyl confirmation that alternates between *anti* and *syn*. The four-stranded telomeric DNA is very stable.

Another important non-covalent mechanism of ligand binding is intercalation. It is occurred when a planner, aromatic molecule slides between the stacked base pairs of the hydrophobic interior of a helical DNA. The unique mode of binding is utilized by agents ranging from simple organic molecules, such as ethidium bromide, to natural products like daunomycin that exhibit antitumor activity. It is a two stage process involving an initial diffusion controlled association of the usually positively charged intercalator with the extension of the helix followed by a slower insertion of the planner moiety between the staked base pairs. After binding intercalator causes a structural purturbation in the helical polymer in order to accommodate the new legend between π staked base pairs. This structural distortions leading to distinct change in the hydrodynamic behavior of the DNA and responsible for increase in viscosity. The change in viscosity causes alteration in gel electrophoretic mobility.

Nucleic acids also involve in chemical reaction that leads to the covalant modification of their structures through alkylation and methylation reactions. The agents responsible for such type of DNA or RNA modifications are led to nucleobase selective modification reactions. The agents sometimes also possess antitumor properties or ironically act as carcinogenic or mutagenic substances, for example, chemotherapic treatment of alkylating chemical mitomycin C is believed to function through formation of monofunctional and bifunctional DNA adducts resulting from crosslinked inter and intra strand guanine residues. The methylation of nucleobase is responsible for enzymatic inactivation of segment. For example, in bacterial DNA the process is responsible for immunization of segment from recognition by enzyme restriction endonuclease.

2.3 BACTERIAL DNA EXTRACTION

Material and Instrumentation

- Material *E.coli* (animal tissue/plant tissue)
- Eppendorf
- TE buffer
- Ethanol
- Phenol
- Ultracentrifuge machine
- Proteinase
- Sodium dodecyl sulphate (SDS)
- Sodium Chloride
- Tris HCl
- EDTA
- Sodium acetate

Procedure

Plant tissues → Homogenized in mortal and pestle

Animal tissues → Homogenized in mortal and pestle

E.Coli → Centrifuged 3000 rpm in 5 min. → supernatant is discarded and pellet is processed further. →

↓

: Add Lysis buffer, Tris, EDTA and Sodium chloride
: 5 ml to 10 ml/gm of material
: Some workers suggested addition of SDS but it is not necessary.

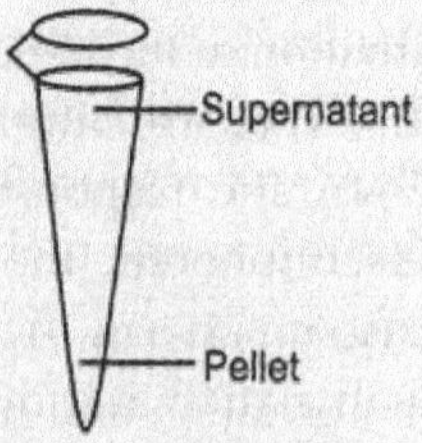

↓

Add 5μl proteinase/trypsin (cellulase/Pectinase for plant cells) from 10 mg/ml of stock solution.

↓

Mix well and incubate at room temperature for 3 min.

↓

Add 1% SDS (10% of stock solution) volume of SDS sol is 50 µl.

↓

Add equal volume of saturated phenol.

↓

Mix for 5 min (very gently otherwise DNAs are broken)

↓

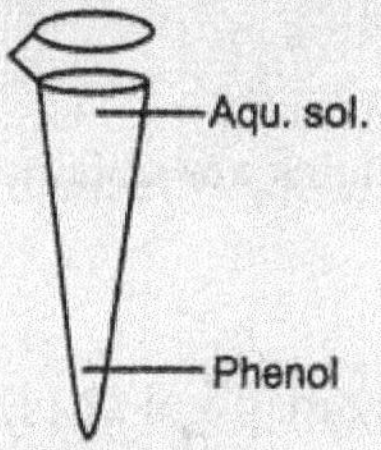

Centrifuged at 3000 rpm for 5 min.

↓

Collect aqueous solution in other eppendorf

↓

Add chloroform and mix gently

↓

Centrifuged at 3000 rpm for 5 min.

↓

Collect aqueous sol. in other eppendorf

↓

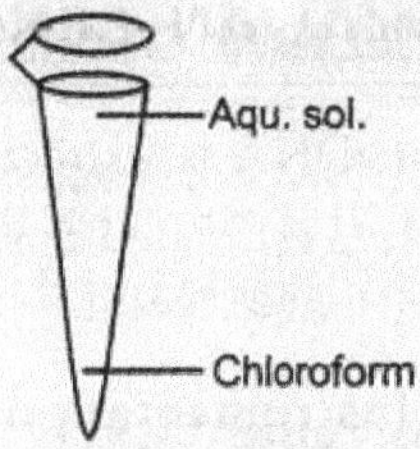

Add equal volume of chloroform and mix
Centrifuged at 3000 rpm for 5 min.

↓

Collect aqueous phase in other eppendorf.

↓

Add 1/10 vol 3M Sodium acetate and mix.

↓

Centrifuged at 3000 rpm for 5 min.

↓

Add 2 vol of ethanol

↓

DNA pelletes are obtained. Discard supernatant and dry. (wet nature is necessary)

↓

Put small vol (2 μl) of sterilized distilled water and keep in eppendorf.

The extracted DNA can be analyzed by gel electrophoresis or used for the preparation of recombinant DNA.

Solution Preparation

1. **Lysis Buffer**
   - NaCl 150 mM
   - Tris. HCl 20 mM pH 8
   - EDTA 50 mM
   - SDS 1%
2. **TE Buffer**
   - Tris. HCl 10 mM pH 8.0
   - EDTA 1 mM

2.4 MAMMALIAN GENOMIC DNA EXTRACTION

The mammalian DNA is extracted from both tissue and blood but blood is preferred for the laboratory work. It is estimated that the yields of DNA vary between 0.5–3.0 μg/mg of tissue and 5–15 μg/300 μl of whole blood.

Material and Instrumentation

- Cell lysis buffer
- Whole blood
- Ethanol
- Isopropanol
- Potassium acetate solution
- Red blood cell lysis buffer
- TE (pH = 7.6)

- DNase free RNase
- Proteinase K
- Water bath (55°C–65°C)
- Eppendorf

Procedure

Take 300 µl aliquots of whole blood in each of two eppendorf

↓

Add 900 µl of red blood cell lysis buffer to each sample and mix gently. Incubate the solution at room temperature for 10 minutes.

↓

Centrifuge the tube at maximum speed for 20 minutes.

↓

Discard all except 20 µl of supernatant.

↓

Resuspend the pellets of white cells in the small volume of supernatant left in each tube. Keep the pellets of both eppendorf into one.

↓

Add 600 µl ice cold cell lysis buffer and homogenize the suspension by strong and forceful mixing. (sometimes a cloudy appearance of solution is observed but it has no effect in extraction of DNA)

↓

Add 3 µl of 4 mg/ml DNase free RNase. Incubate for 15–60 minutes at 37°C

↓

Cool at room temperature and add 200 µl of potassium acetate solution. Mix the contents of the tube by vortexing vigorously for 20 seconds.

↓

Centrifuge at maximum speed for three minutes at 4°C. A pellet of protein should be visible at the bottom of the eppendorf after centrifugation. If not incubate the lysate for five minutes on ice and repeat the centrifugation step.

↓

Take other eppendorf with 600 µl isopropanol, collect supernatant and mix well gently.

↓

Centrifuge the eppendorf at maximum speed for one minute at room temperature. DNA is precipitated.

↓

Discard the supernatant and add 600 μl of 70% ethanol to the DNA pellet. Mix by inversion of eppendorf and centrifuge at maximum speed for one minute.

↓

Carefully remove the supernatant and allow the DNA pellet to dry in air for 15 minutes.

↓

Redissolve the pellet of DNA in 100 μl TE (pH = 7.6). The solublization of the genomic DNA can be facilitated by incubation at 60°C.

Solution Preparation

1. Red blood cell lysis buffer
   20 mM TRIS-Cl (pH = 7.6)
2. Cell lysis buffer
   10 mM Tris-Cl (pH = 8.0)
   1 mM EDTA (pH = 8.0)
   0.1% (w/v) SDS
   Store the buffer at room temperature but chill an aliquot to 0° before use.
3. Potassium acetate solution
   5M potassium acetate = 60 ml
   Glacial acetic acid = 11.5 ml
   Distilled water = 28.5 ml
   The solution is 3M with respect to potassium and 5M with respect to acetate.

2.5 YEAST DNA EXTRACTION

Material and Instrumentation

- Ethanol
- Phenol: Chloroform (1:1 v/v)
- Sodium acetate (3M, pH = 5.2)
- STES buffer
- TE (pH = 7.6)
- Acid washed glass beads (0.4 mm)
- Yeast cell
- YPD medium

Procedure

Inoculate a yeast colony into 10 ml of YPD medium and incubate over night at 30°C.

↓

Take 1.5 ml from an overnight culture of yeast cell in an eppendorf.

↓

Centrifugate at 4000 rpm for one minute

↓

Remove the culture medium and resuspend the pellet in 50 µl of STES buffer.

Colonies on plate

↓

Use loop and transfer one large colony in an eppendorf

↓

Add 50 µl of acid washed glass beads to each tube containing the resuspended yeast.

↓

Add 20 µl of TE (pH = 7.6)

↓

Add 60 µl of Phenol: Chloroform (1:1 v/v) and mix well by vortexing for one minute.

↓

Centrifugate at 4000 rpm for 5 minutes.

↓

Transfer the upper aqueous phase to a fresh eppendorf tube. Add ethanol and keep it for 20 minute

↓

Centrifuged at 4000 rpm for 10 min at 4°C

↓

Remove supernatant and rinse the pellet with 100 µl of 70% ethanol

↓

Centrifuged at 4000 rpm for one min.

↓

Discard supernatant and keep pellet (DNA) dry in the air for 15 min.

↓

Add 40 µl of TE (pH = 7.6)

Preparation of Solution

1. STES buffer
2. 0.2 M Tris-Cl
3. 0.5 M NaCl
4. 0.1 % (w/v) SDS
5. 0.01 M EDTA

2.6 ESTIMATION OF DNA

In the electrophoresis the staining dye used for the observation of DNA is ethidium bromide. Since the amount of fluorescence emitted by the ethidium bromide bound to DNA is proportional to the total mass of DNA in a band on a gel can be estimated by comparing its intensity of staining with that of a DNA mass standard. The amount of DNA present in the sample could also be estimated by taking absorbance reading at 260 nm and 280 nm. Nucleic acid, including DNA, absorbs maximally at 260 nm. Protein absorbs maximally at 280 nm. The ratio between reading at 260 nm and 280 nm provides an estimate of purity of the sample. Pure preparation of nucleic acids have A260/A280 ratios of 1.8 to 2.0, if the sample contains protein or other contaminants then the ratio is much lower and it is not possible to accurately quantitate the amount of DNA with this method. Quantity of DNA is estimated by

1. Colorimetric estimation.
2. UV–VIS spectrophotometer method.
   (*i*) It is color dependent estimation in which monochromatic light is created by the use of filter.
   (*ii*) The monochromatic light is created by prism. The prism is made of quartz or silica. The source of UV light is Xenone.

The principle of coloration is DNA reacts with diphenylamide (1 ml)to give blue green color. After mixing DNA with reagent the solution is boiled for 2 min and cool. In colorimeter estimation the filter color is determined by the complementary color of solution to be tested. As the complementary color of blue green is red, we use red filter to findout OD (–1 μg). DPA is prepared in sulphuric acid. In spectrophotometer very small amount of DNA can be estimated (μg). In the instrument colorless sample is used.

DNA → Sample 1, Sample 2 + Blank and standard (calf thymine DNA) (10 mg/ml)

Procedure

1. Pipette out different volumes of DNA solution into test tubes.
2. Make the volume to 3 ml with water.
3. Prepare the fresh reagent by dissolving 1 g of diphenylamine in 100 ml glacial acetic acid and by adding 2.5 ml of conc. sulphuric acid.
4. Add 5 ml of the reagent, mix well and heat in a boiling water bath for 10 minutes.
5. After cooling measure take the optical density with a red filter (or 595 nm).

DNA (5 µl)

↓

1 µl DNA

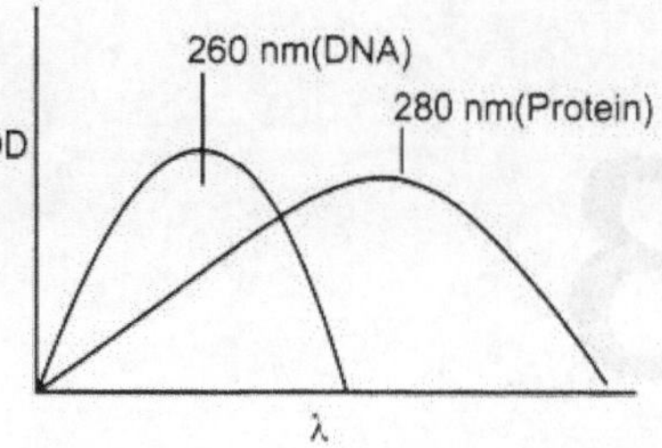

Calculation

(A) DNA ⟶ 20 µl

↓

5 µl ⟶ 1 µl

OD_{260} = 0.1 observed ratio

OD_{280} = 0.05 2:1 = DNA : Protein

If ratio comes 1:1 represents more protein and requires further phenolic extraction .

Dilution factor = 1000 µl/5 µl = 200 times

OD_{260} × DF = OD/ml

0.1 × 200 = 20 OD/ml

1 OD = 50 µg (standard for DNA)

1 OD = 40 µg (standard for RNA)

1 OD = 50 µg

20 OD = 50 × 20 = 1000 µg/ml

1000 µl = 1000 µl/ml

20 µl = 1000/1000 × 20 = 20 µg DNA

(B) DNA = 50 µl

↓

10 µl ⟶ 1000 ml

OD_{260} = 0.8 | observed

OD_{280} = 0.5 |

Dilution factor = 1000/10 = 100

OD_{260} × DF = 0.8 × 100 = 80 OD/ml

10 D = 50 µg

80 OD = 50 × 89 = 4000 µg/ml

1000 µl = 4000 µg

50 µl = 4000/1000 × 50 = 200 µg DNA

Chapter 3

EXTRACTION OF RNA

3.1 INTRODUCTION

There are different species of RNA present in the cell. They are classified according to their functions or by size (molecular weight or sedimentation coefficient). The three main types of RNA according to their functions are ribosomal, transfer and messenger RNA (*r*-RNA, *t*-RNA and *m*-RNA respectively). Ribosomal RNA molecules are structural components of ribosome and constitute the bulk of cellular RNA (approximately 75%). Each ribosome contains three different molecules of *r*-RNA classified by size: 28S, 18S and 5.5S in eukaryotic cytoplasmic ribosomes or 23S, 16S and 5S in prokaryotic ribosomes. The transfer RNAs are soluble in cytoplasm and constitutes 15–18% of total RNA present in cell. Prior to the formation of the initiation complex during protein synthesis the families of *t*-RNA are activated or acylated for carrying of desired amino acids from cytoplasm to the site of protein synthesis necessary for the elongation of polypeptide chain. The messenger RNA is heterogenous and constituted 3–7% of total RNA. It forms the negative strand for polypeptide chain and determines the sequence of amino acids join successively in chain during protein biosynthesis.

Total nucleic acid can be extracted from the tissue, followed by removal of DNA and purification of the RNA. This is the usual method for the isolation of large quantities of *t*-RNA or total *m*-RNA. The alternative procedure is to begin the isolation with subcellular fractionation. This method is commonly used for the isolation of specific *m*-RNA or for isolation of RNA from specific organelles (*e.g.*, chloroplasts and mitochondria). Transfer RNA can be isolated successfully by either one of two methods stated above.

RNA extraction should be done with extreme care to avoid ribonuclease activity because ribonucleic acids are enzymes, and are active even in minute quantities.This can be prevented by using any one of the ribonuclease inhibitors such as heparin placental RNasin, and Diethyl pyrocarbonate (DEPC) or by using guanidium salt in extraction buffers.

Method of RNA Extraction

There are several methods for extraction of RNA from plants, animal cells and tissues. The selection of particular method depends on the precise requirements of the experiments, the biological material used for extraction. The selection of method is also determined by the type of RNA required for study or purification. The methods commonly used for extraction of RNA are as follows.

1. *Phenol Method :* In this method either hot phenol or cold phenol is used as deproteinising agent during the extraction procedures.
2. *Guanidium Salt Method :* It avoids the use of phenol but high molarity of guanidium salt is used to dissociate the protein from RNA. The guanidium salt also acts as inhibitor of ribonuclease activity and there by helps in extraction of intact RNA.
3. *Lithium Chloride or Urea Method :* It is useful in preparing *m*-RNA from variety of sources.

3.2 CHEMISTRY OF RNA

Chemically RNA is polynucleotide, linked by phosphodiester like DNA but it contains D ribose sugar ring in its chain and an aromatic nucleobase. Chain has a constant 5′ to 3′ polarity with negatively charged sugar—phosphate backbone and an array of relatively hydrophobic nucleobase. The nitrogenous bases represent amphiphilic features which ultimately drive the assembly and maintenance of 2° and 3° nucleic acid structure. It has been suggested that C_2—OH substituents of RNA causes this nucleic acid to adopt configurations that differ from those of DNA (*Saenger*, 1984). The post transcriptional modifications of nucleobase are also observed particularly with *t*-RNA (*Hall* 1971). These modified nucleobase, which are thought to "tune" the stability and three-dimensional shape of the nucleic acid to suit particular biological roles.

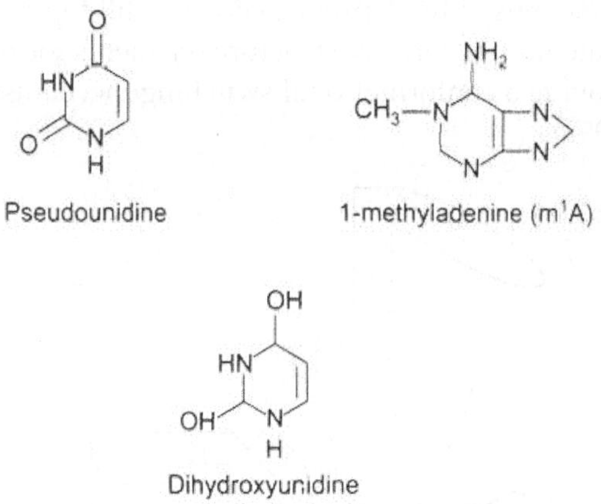

FIGURE 3.1. *Represents nitrogenous bases modified after transcription*

The current research publications supported the idea that the precise function of an RNA molecule relies on its ability to maintain a proper three-dimensional structure (*Altman*, 1984, *Cech*, 1987, *Rould et al* 1989). The *t*-RNA is smallest and consists by 60–75 ribonucleotides. It represents unusual base pairing (G:U) in acceptor stem. The 2° configuration is characterized by presence of four stems and four loops. The anticodon loop is responsible for degeneracy of triplet code due to presence of wobble position. The 3° configuration of *t*-RNA is L-shaped.

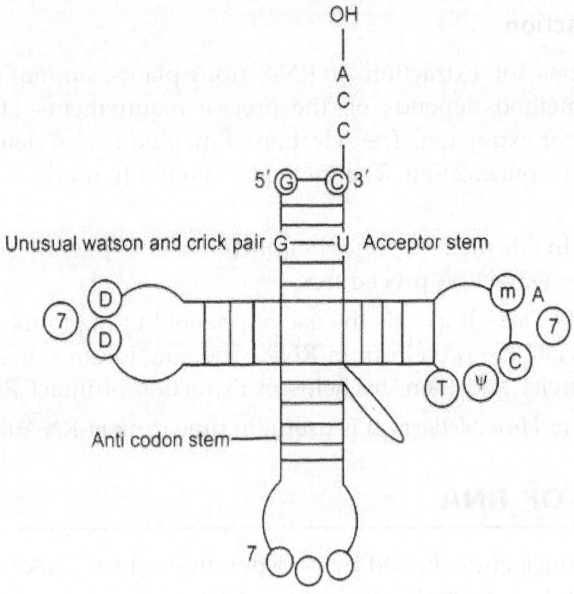

FIGURE 3.2. *Represents clover leaf (2°) configuration of t-RNA*

The ribosome contains 2/3 rd part RNA of its gross weight whereas 1/3 rd part is of proteins. The number of ribonucleotides for 5s *r*-RNA, 16s *r*-RNA and 23s *r*-RNA are 120, 1542 and 2904 in numbers respectively. They incorporate many of the elements of 2° structure represented by *t*-RNA molecules *i.e.*, array of double stranded stems and single stranded loops. In addition there are interior loops, bulge loop and multiple branched loop structure. There is increasing evidence to suggest that *r*-RNAs often contain a structural unit termed the pseudoknot (*Dam et al* 1992, *Schimmel*, 1989). The exact functions of RNA pseudoknot are still topic of active investigation. But it is speculated that they create unique tertiary structural elements for recognition by other RNAs or proteins and may also provide a conformational switching mechanism between two structural forms of RNA (*Dam et al* 1992).

FIGURE 3.3. *Pseudoknot (a part of) r-RNA 2° configuration*

Although *m*-RNA is single stranded, the hairpin or stem loop structures are present. These specialized configurational structures are provided a means of controlling the rate of expression of a particular gene by providing stopping and pausing points of the ribosomal machinery. Such regions are called terminator or anti-terminators. The process has been studied within the trp-operon and at least the two regions are recognized which is responsible for the development of ability of RNA to actively participate in controlling its own rate of transcription and translation (*Yanofsky*, 1988).

The varieties of structural features and chemical functional groups found in nucleic acids enable them to interact with many different molecules by any of three methods.

1. Non-covalent interaction
2. Alkylation
3. Polymer strand scission

1. Non-Covalent Interaction

It is most important chemical method in promoting the molecular recognition of nucleic acids. The nucleobase heteropolymers and the anionic phosphates participate in the formation of hydrogen bond, *van der Waals* interaction, hydrophobic contents and salt bridge. In addition hydrophobic surfaces of the tertiary fold assist in the binding and orientation or ligands. One of the key participant of non-covalent interaction is the negatively charged ribose phosphate backbone. The anionic phosphate of the backbone interect electrostatistically with cations such as Na^+ or Mg^+, increasing the structural stability of the polymer by screening the phosphate-phosphate repulsive interactions, particularly with folded RNA.The polyanionic nature of nucleic acids is also essential during the initial stages of the binding of positively charged proteins or small molecules.

Once the nucleic acid and potential binding ligand are in close proximity, short range, weak molecular forces begin to participate in their interactions. These weak interactions are usually combinations of hydrogen bonds with the functional groups of the nucleobase, hydrophobic interactions and salt bridge between a positively and negatively charged moiety of the binding ligands and an anionic phosphate, with protein that bind to RNA. These interactions usually involve the side chains of amino acids which can act as hydrogen bond donars or acceptors provide sites of hydrophobicity or participate in ionic interactions. Specifically, in the case of *t*-RNA these features permit the correct recognition of an amino acid by the aminoacyle—*t*-RNA synthetase.

Intercalation

Another important non-covalent interaction exists between ligand and nucleic acid is intercalation (*Waring*,1970). It develops when a planner and aromatic molecule slides between the stacks of base pairs of the hydrophobic interior. While most of the knowledge are confinded with the study of DNA , the tertiary structure of RNA also appears to enable ligand to bind through this mechanism (*Chow* and *Barton*, 1990). Intercalation is a two stage process involving :

(*a*) An initial diffusion controlled association of the usually positively charged intercalator with the exterior of the helix.

(*b*) A slower insertion of the planner moiety between the stacked base pairs (*Long* and *Barton*, 1990).

2. Alkylation

Nucleic acid also participates in chemical reactions that result in the covalent modification of their structures through alkylation or metallation reactions. Agents that lead to RNA modifications can often function in nucleobase selective modification reactions. Such agents sometimes also possess antitumour properties or act as carcinogenic or mutagenic substances. For example, organic reagent dimethyle sulfate acts as selective modifier of guanine nucleobase and converts it to N_7-methylated guanine. Thereafter N_7-methylated guanine creates positive charge to the heterocycle and promotes the hydrolysis of the N-glycisidic bond. The resultant is responsible for the formation of an apurinic or abasic site, (lost purine without breaking of phosphoester bond).

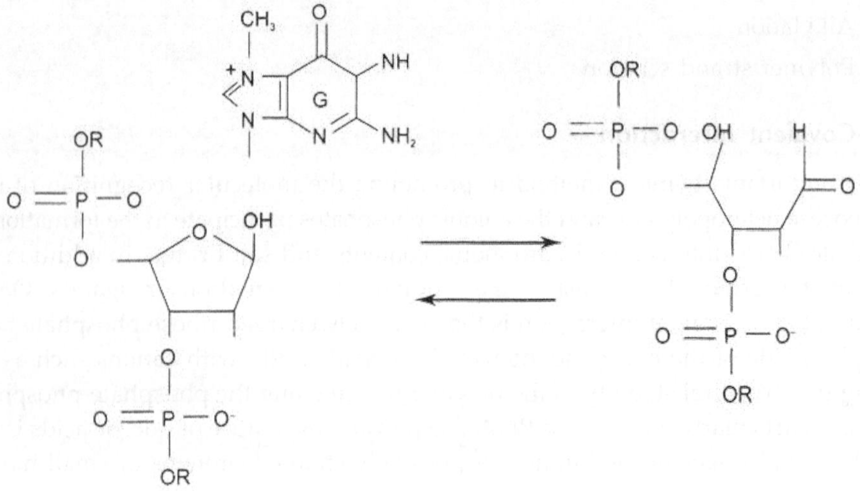

3. Polymer Strand Scission

Nucleic acids are also chemically altered through strand scission of their polymeric backbone. This type of nucleic acid modification can occured through several mechanisms. But only three important methods are described in the present book.

(*a*) Oxidation of the ribose sugar ring

(*b*) Hydrolysis of phosphodiester bond

(*c*) RNA self-processing reaction.

(a) Oxidation of the Ribose Sugar Ring

The oxidative degradation of the sugar moiety is caused due to homolytic or heterolytic breakage of a C-H bond of the ring. It results the formation of a carbon centered radical or cations; leads to intramolecular reorganization and breaking of phosphoester bond. For example, Febleomycine and dynemycine are responsible for strand scission by oxidizing the ribose sugar moiety. The chemical steps are as follows:

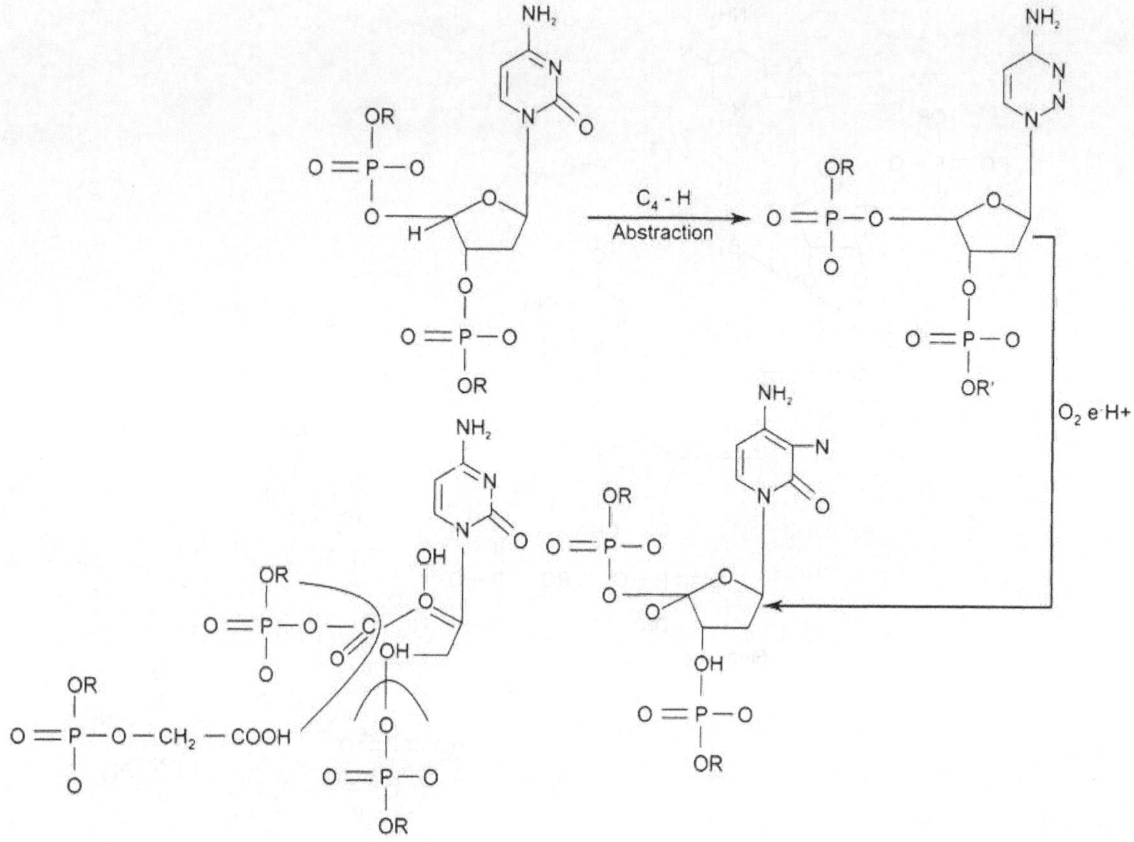

(b) Base Protonated–hydrolysis in RNA

The base protonated hydrolysis is initiated when a chemical acts as Lewis base and donates proton, the common example is Fe^{2+} - EDTA. It donates proton to 2'-OH of ribose sugar, responsible for intra-chain re-organization due to sequential steps acts as Lewis acid and Lewis base. The sequential reorganization reaction finally breaks phosphoester bond as illustrated below. The hydrolysis of the DNA backbone by this method due to absence of 2'-OH group in ribose sugar ring has been much more difficult to achieve. As in ribose sugar ring of RNA 2'-OH functional group is not available it can assist in the cleavage of phosphodiester bonds between ribonucleotides.

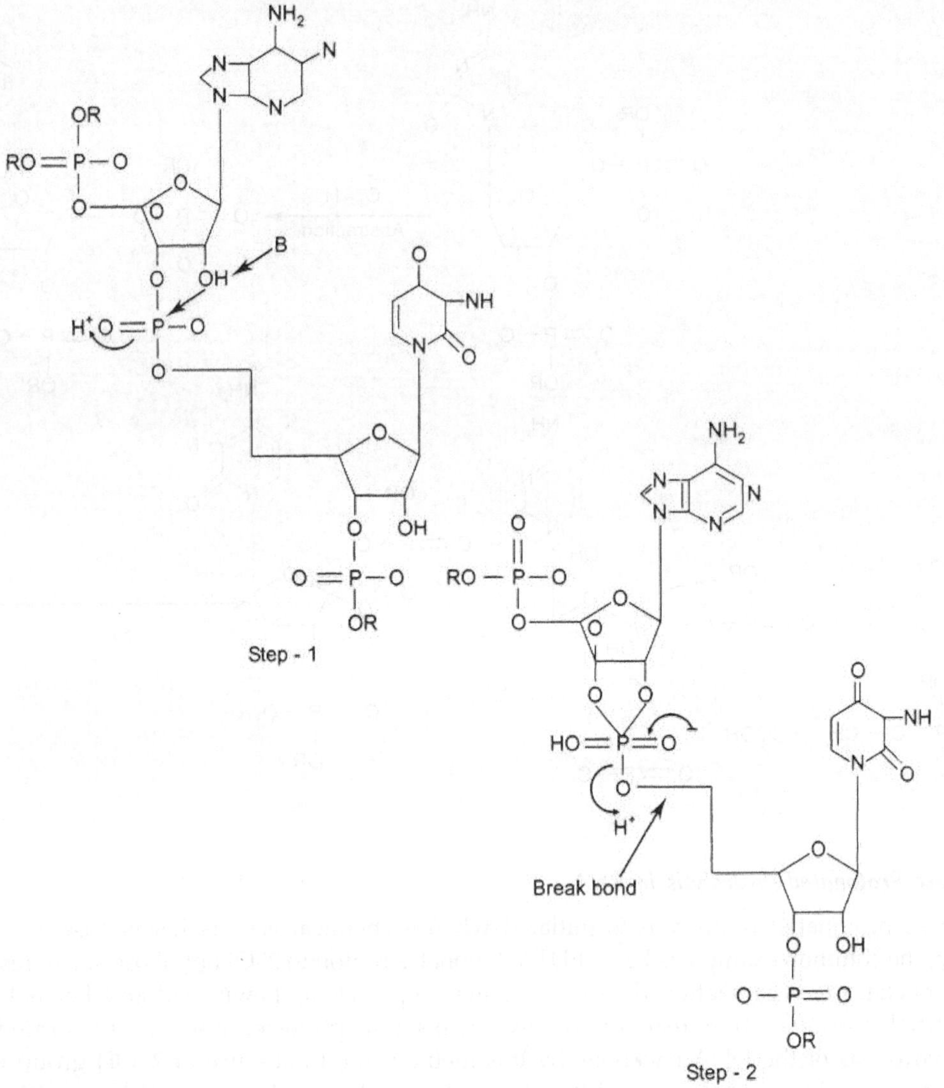

(c) RNA Self-Processing Reaction

The both methods of strand scission, described above, are caused by exogenous reagent. But some RNAs are also capable of affecting their own modifications through self-processing reactions. While several examples of self-processing RNAs have been reported, perhaps the best known are the intervening sequence of the *Tetrahymena thermophilia* pre-*r*-RNA and the M1 RNA componant of RNase P.

The enzymes (ribozymes) catalyse the process leads to removal of 413 nucleotide present at the 5' terminus of *t*-RNA precursor. The autocatalytic reaction of ribozyme is as follows :

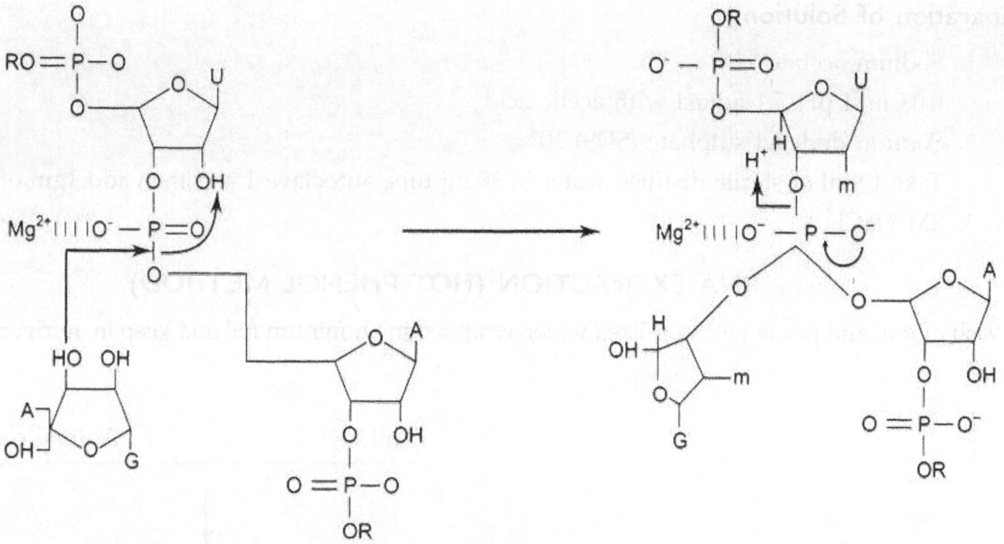

It is a reaction associated with a series of transesterification that are initiated by the free 3′OH or of a guanosine nucleotide cofactor. In this process 3° configuration of RNA also plays a significance role.

3.3 METHOD FOR EXTRACTION

1. Wash the cells with saline (0.15 M NaCl).
2. Lyse in 10 volumes of 0.15 M acetate buffer pH 5.1 and SDS to a final concentration of 1.0%, Heparin 2 ug/ml. Mix gently and leave at room temperature for 2 min.
3. Add one volume of hot acid phenol 80% at 65°C.
4. Incubate at 65°C for 10 min.with intermittent shaking.
5. Centrifuge at 10,000 rpm for 15 min.
6. Remove the aqueous phase using a Pasteur pipettes and saves in ice.
7. To the organic phase add equal volume of 0.05 M acetate buffer.
8. Mix thoroughly and centrifuge at 10,000 rpm for 5 min.
9. Remove the aqueous phase.
10. Pool aqueous phases and do chloroform : isoamyl alcohol extraction.
11. Collect aqueous phase. Add 0.17 M NaCl or volume 3 M Sodium acetate pH 5.1 and mix.
12. Add 2.5 volume of ice cold ethanol, store at –20°C for 24 hours.
13. Pellet at 12,000 rpm for 10 min.
14. Dissolve the pellet in sterile distilled water and measure O.D. at 260 and 280 nm. Or estimate by orcinol method.
15. Store at –20°C (for immediate use) or store in ethanol 2.5 volume in the presence of 0.17 M NaCl with 0.05% diethylpyrocarbonate (for long time storage).

Preparation of Solution

1. Sodium acetate
   0.05 mM pH 5.1 adjust with acetic acid
2. Sodium dodecyl sulphate (SDS) 10%.
   Take 10 ml of sterile distilled water in 30 ml tube autoclave 1 and then add 1gm of SDS.
3. 3M NaCl.

RNA EXTRACTION (HOT PHENOL METHOD)

Wash mortal and pestle with sterilized water, wrapped in aluminum foil and keep in refrigeration.

E.Coli

↓

Plant cell | Animal cell

↓

Centrifuged 3000 rpm for 5 min — Surface sterilized by alcohol

↓

Wash in distilled water

↓

0.05 M Sod. Acetate (500 μl) ← Homogenized in 0.05 M. Sod.acetate

↓

Add lysozyme + trypsin + protease + cellulase only for plants (20 μl),

↓

Mix and incubate at room temperature for 5 min.

↓

Add 50 μl 10% SDS and mix; incubate for 5 min. at room temperature.

↓

Add equal volume of saturated phenol.

↓

Heat 55–65°C in water bath for 5 min.

↓

Centrifuged 3000 rpm for 5 min.

↓

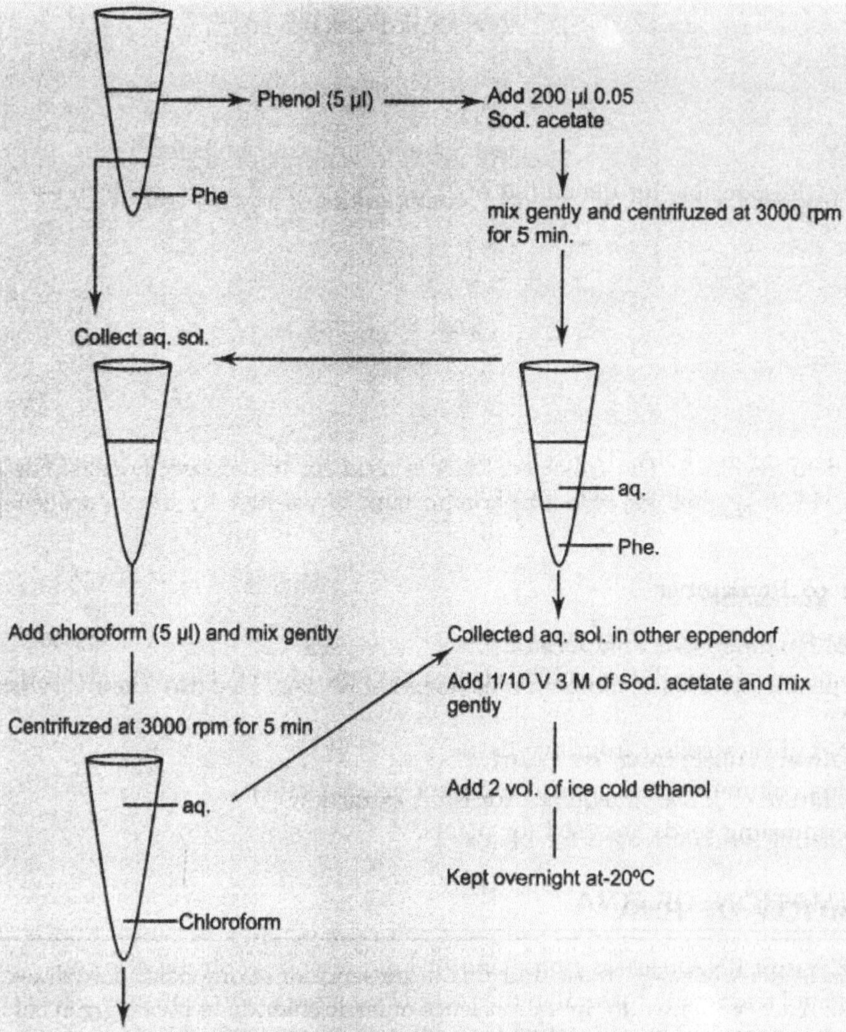

–20°C overnight

↓

Centrifuged at 5000 rpm for 5 min.

↓

Dry pellets

↓

Wash with 70% ethanol

↓

Dissolve 10 μl distilled water

↓

Heat 55°–65°C in water both for 5 min.
(Responsible for denatured of 2° configuration if any present in sol.)

↓

Cool for 1 min.

↓

Centrifuged to get RNA. The collected RNA is used for the electrophoresis. For RNA we commonly use 1.5% agarose gel. Electrophoretic tank is washed by 5% hydrogen peroxide overnight.

Important to Remember

1. Careful handing is very important.
2. Some people use RNase, inhibitors like placenta RNasin, Heparin, Diethylpyrocarbonate etc.
3. RNA always collect from live tissue.
4. 3 hrs culture of *E.Coli* is required for RNA extraction.
5. Germinating seeds are used for plants.

3.4 ESTIMATION OF RNA

Acid hydrolysis of RNA releases ribose and this in presence of strong acid, dehydrates to yield furfural. Orcinol reacts with furfural in the presence of ferric chloride to give a green color, Purine nucleotides are generally more reactive than pyrimidine nucleotides.

Procedure

1. Set up different tubes with increasing concentrations of RNA.
2. Make up the volume to 2.0 ml with water.
3. Prepare orcinol reagent by dissolving 300 mg of orcinol in 5 ml of ethanol.
4. Add 3.5 ml of this just before use to 0.1 % solution of ferric chloride in conc. HCL.
5. Add 3 ml of orcinol reagent to each tube and mix well.
6. Keep the tubes in a boiling water bath for 15 min. after which cool them to room temperature.
7. Measure the O.D using a red filter (or 665 nm).

Estimation of RNA

1g ⟶ 10 µl

↓

1 µl ⟶ 1 ml (1000 µl)

OD_{260} = 0.05

OD_{280} = 0.01

Dilution factor = 1000/1 = 1000

DF × OD_{260} = 1000 × 0.05 = 50 µg/ml

1 OD = 40 µg

50 × 40 = 20,00 µg/ml

1000 = 2000 µg/ml

10 µl = 2000/1000 × 10 = 20 µg RNA

Estimation of Protein

1 g ⟶ 10 ml ⟶ 20 ml ⟶ 1 ml.

OD_{260} = 0.5 observed

OD_{280} = 1.0

Dilution factor = 1000/20 = 50 times

1.55 × OD_{280} – (0.75 × OD_{260}) = mg/ml Protein

1.55 × 1 – (0.75 × 0.5) = 1.55 – 0.375

= 1.175 × 50 = 58.75 mg/ml

= 58.75 × 10 = 587.5 mg

For RNA

20 µl ⟶ 2 µl -1ml

OD_{260} = 0.1

DF = 1000/2 = 500

0.1 × 500 = 50 OD

1 OD = 40 µg

50 OD = 40 × 50 = 2000 µl/ml

20 µl = 2000/1000 × 20 = 40 µg/ml

For DNA

50 µl ⟶ 5 µl ⟶ 1 ml

OD_{260} = 1

DF = 1000/5 = 200

1 OD = 50 µg

$$200 \times 50 = 10{,}000\ \mu g/ml$$
$$1000\ \mu l = 10{,}000$$
$$50\ \mu l = 10{,}000/1000 \times 50 = 500\ \mu g/ml$$

1 g tissue now has

$$\text{Protein} = 587.5\ mg$$
$$\text{DNA} = 1000\ \mu g$$
$$\text{RNA} = 80\ \mu g$$

Specific activity

$$587.5\ mg = 1000\ \mu g\ \text{DNA}$$
$$1\ mg = 1000/587.5 = 1.8\ \mu g\ \text{DNA/mg protein}$$
$$587.5\ mg = 80\ \mu g\ \text{RNA}$$
$$1\ mg = 80/587.5 = 0.136\ \mu g\ \text{RNA/mg protein.}$$

3.5 SUB-CELLULAR FRACTIONATION

The unit of life is cell and it contains specialized intracellular (some times extra cellular also like cilia and flagella) structures for different functions, known as organelles. Some times it is necessary to isolate the cell organelles for the study of biochemical of metabolic activities of different cell organelles. The purity of extraction is also fundamental necessity of the work. The isolation of a particular cell organelle in large quantity is generally possible by technique of sub-cellular fractionation or differential fractionation. The basic principle for the method is stated as " the cell organelles or particles of different size and shape travel towards bottom of a centrifuge tube at different rates when placed in a centrifugal field, as long as they are more dense than the surrounding medium. The speed of ultracentrifugation is represented in rpm (revolution per minute) and force applied in ultracentrifugation is represented in *g*. The velocity of a given particle during centrifugation in a fluid medium is directly proportional to the square of the angular velocity (ω) and the distance of the particle from the center of the rotor (r).

$$V = s\,(\omega^2 r)$$
$$g = (\omega^2 r)$$

where, ($\omega^2 r$) is termed as radial acceleration and represented in centimeter per second square and s is termed as sedimentation co-efficient and it is equivalent to the average velocity per unit acceleration. The notation g represents centrifugal acceleration and is generally expressed relative to earth gravitational acceleration (910 cm/sec^2).

For sub-cellular fractionation, at first cells are broken commonly by mechanical disruption in an isotonic buffered solution using mechanical homogenizer. The homogenate is then processed by a series of sequential centrifugation. Initially, the homogenate is subjected to low centrifugal forces for a short period of time so that only the largest of the cellular organelles are sedimented into a pellet. With successively greater centrifugal forces larger cytoplasmic organelles can be spun out of suspension, then the microsomes and finally the ribosome.

Once ribosome has been removed the supernatant consists of the cell soluble phase and those particles too small to be readily removed by sedimentation. The initial steps of differential centrifugation do not generally yield pure preparation of particular organelles; so further steps are usually required. In many case, further purification is accomplished by centrifugation of the crude preparation through a sucrose density gradient. Centrifugation through gradient distributes the contents of the fraction into various layers according to the density of the components.

Material and Instrumentation

- Living tissues
- TNM buffer
- Sucrose gradient
- Homogenizer
- Utracentrifugation
- Eppendorf

Process for Laboratory

Tissue/Cells

Homogenized in TNM buffer (5–10 ml/gm)

↓

Collect some filtrate for protein estimation

↓

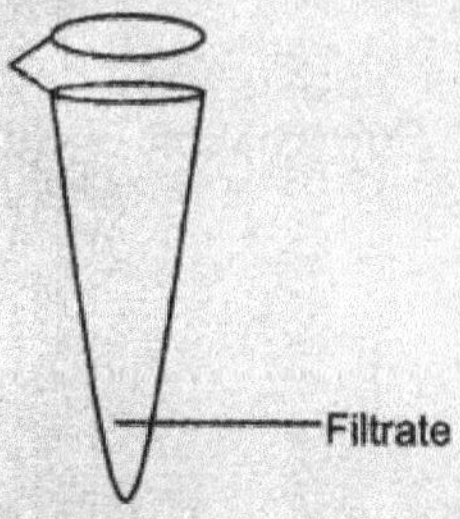

Filtered and the filtrate is collected in eppendorf

↓

Centrifuge 3000 rpm for 5 min

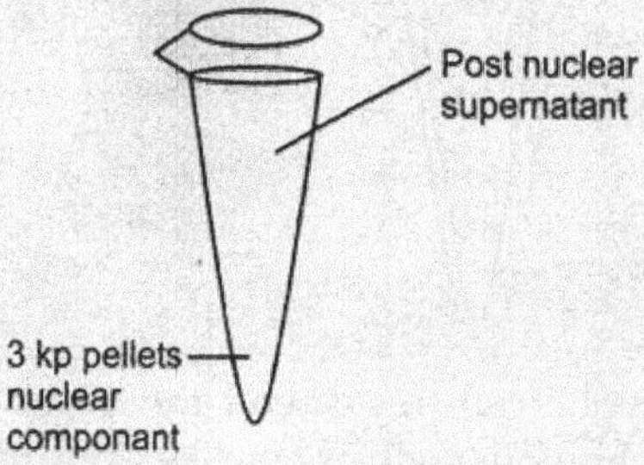

↓

Centrifuged at 10,000 rpm for 1/2 h

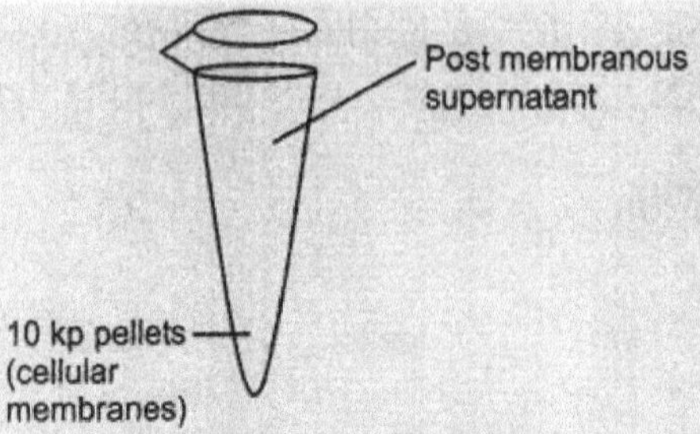

Post membrane Supernatant is collected in an eppendorf.

↓

Centrifuged at 28,000 rpm for 1 h

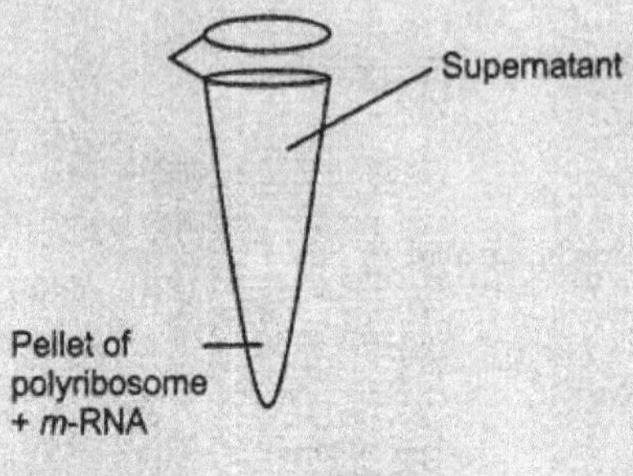

Supernatant is collected in eppendorf

↓

EDTA Treatment

↓

Centrifuged at 60,000 rpm for 4 h

↓

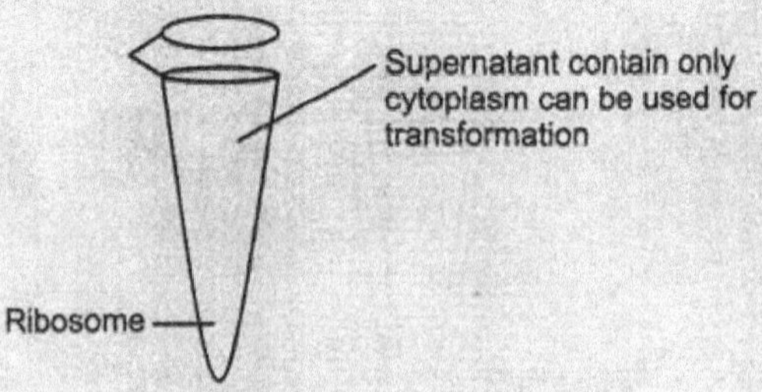

In the condition of eukaryotic cell, resuspend the 10 kb pellets for further differentiatial fractionation to separate membranous cell organelles. The material is collected in eppendorf containing suitable liquid medium (1 M sucrose solution). The eppendorf is centrifuged at 25000 rpm for one hour. After the process keep the solution still for sometime. The solution represents three bands for uppermost lysosome, middle mitochondria and lowermost micro bodies.

Chapter **4**

Extraction of Plasmid DNA

4.1 INTRODUCTION

Plasmids are circular, self-replicating extra chromosomal genetic material of bacteria. The same bacterial enzymes used for duplication of bacterial chromosome carry out replication of plasmid. DNA replication originates at a specific sequence called the origin of replication. It is commonly used as vector. Plasmid cloning vectors have been engineered to contain a region of closely arranged restriction enzyme cleavage sites for easy insertion of foreign DNA. This region is called the multiple cloning region (MCR). This region contains multiple restriction enzyme cleavage sites so that a person can choose among a variety of restriction enzymes for cloning.

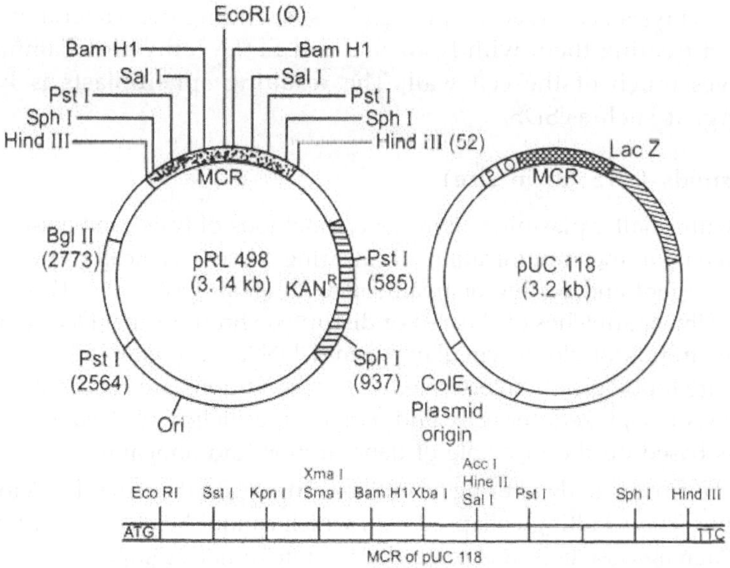

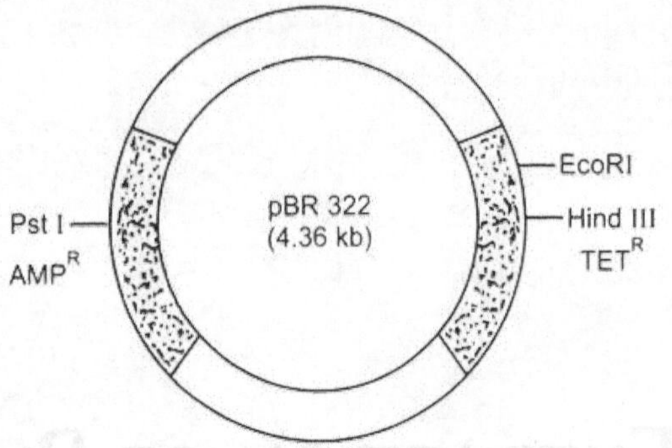

FIGURE 4.1. *Represents the plasmid vectors pRL 498, pBR 322 and pUC 118 and their recognition sites for different endonucleases.*

Whenever possible, plasmids should be purified from bacterial cultures that have been inoculated with a single transformed colony picked from an agar plate. Usually, the colony is transferred to a small starter culture, which is grown to late log phage (Chapter-15). Aliquots of this culture can be used to prepare small amounts of the plasmid DNA (minipreparation) for analysis and/or as the inoculums for a large scale culture.

Bacteria are recovered by centrifugation and lysed by any one of the methods including treatment with non-ionic or ionic detergents, organic solvents, alkali and heat. The choice among these methods is determined by three factors; the size of the plasmid, the strain of *E.coli* and the technique used subsequently to purify the plasmid DNA. The following general guidelines can be used to choose a method that will give satisfactory result.

Large Plasmid (>15 kb in size)

Plasmids >15 kb in size are susceptible to damage during both cell lysis and subsequent handling. Gentle lysis is best, accomplished by suspending the bacteria in an isotonic solution of sucrose and treating them with lysozyme and EDTA (ethylene diamine tetra acetic acid), which removes much of the cell wall. The resulting spheroplasts is lysed by adding an anionic detergent such as SDS.

Smaller Plasmids (<15 kb in size)

When work with smaller plasmids, more severe methods of lysis can be used, and no special care should be taken for the minimization of shearing forces. Typically, bacterial suspensions are exposed to detergent and boiling or treatment with alkali lyses cells. This disrupts base pairing and causes the linear stretches of sheared or disrupted chromosomal DNA of the host to denature. However, the strands of closed circular plasmid DNA are unable to separate from each other because they are topologically intertwined. When conditions are returned to normal, the strands of plasmid DNA recognized perfectly and native super helical molecules are reformed. Basically the method is based on the principle of denaturation and annealing.

Prolonged exposure to denaturing conditions causes closed circular DNA to enter an irreversible denatured state. The resulting collapsed coil, which cannot be cleaved with restriction enzymes, migrates through agarose gels at about twice the rate of native super helical closed circular DNA and stains poorly with intercalating dyes such as ethidium bromide. Varying amounts of this collapsed form of DNA can usually be seen in plasmids prepared by alkaline or thermal lysis of bacteria.

All methods of lysis yield preparation of plasmid DNA that are always contaminated with considerable quantities of RNA and variable amounts of genomic DNA. Crude preparation of plasmid DNA can be easily visualized in agarose gels and can be used as templates and substrates for most restriction enzymes and DNA polymerase. However contaminants must be removed or at least reduced to manageable levels. The purified plasmid is necessary when transfacting mammalian cells.

Of the hundreds of methods of plasmid DNA purification described in literature the alkaline lysis method (*Birnboin* and *Doly* 1979, *Ish-Horowicz* and *Burke*, 1981) is most popular because of its simplicity, relative low cost and reproducibility. It is widely used in most laboratories for minipreparations. For large scale preparation the method of choice is alkaline lysis followed by differential precipitation of plasmid DNA with polyethylene glycol (*Lis*, 1980), which yields contamination free DNA fit for transfection of mammalian cells and all enzymatic reactions, including DNA sequencing.

4.2 METHOD FOR EXTRACTION

Material and Instrumentation

- *E.coli*
- Eppendorf
- Sodium dodecyl sulphate (SDS)
- Bench top centrifuge
- Sodium hydroxide
- Potassium acatate
- Phenol
- Isopropanol
- LB medium
- Ether
- TRIS (pH = 8)
- EDTA
- Glucose

The bacterium *E.coli* is cultured in medium LB | YT | Terrific Broth. It is necessary to maintain following during the process of inoculation for better results.

1. Culture is adequately aerated
2. The volume of the culture tube should be four times greater than the volume of the bacterial culture
3. The tube should be loosely capped
4. The culture should be incubated with vigorous agitation.

Procedure

2 ml LB medium of *E.Coli*

↓

1 ml eppendorf × 2

↓

Centrifuged at 3000 rpm for 5 min. at room temperature

↓

Discard supernatant (remaining contains broken component of microorganism)

↓

Add sol-I (200 μl) contains Tris (10 mM), EDTA (1mM) and glucose (100 mM). Here EDTA acts as metal chelator.

↓

Add sol-II (200 μl) containing freshly prepared 0.2N NaOH and 0. 5% SDS. The solutions are used from stock solutions contain 2N NaOH and 20% SDS.

↓

Incubate for 10 minute at room temperature.

↓

Add sol-III (200 μl)

3M Potassium acetate (pH = 5.1) is responsible for renaturation (establishment of plasmid configuration), known as differential renaturation.

↓

Incubate for 10 min in ice/freeze (do not shake)

↓

Filter to remove chromosomal DNA by the use of cotton.

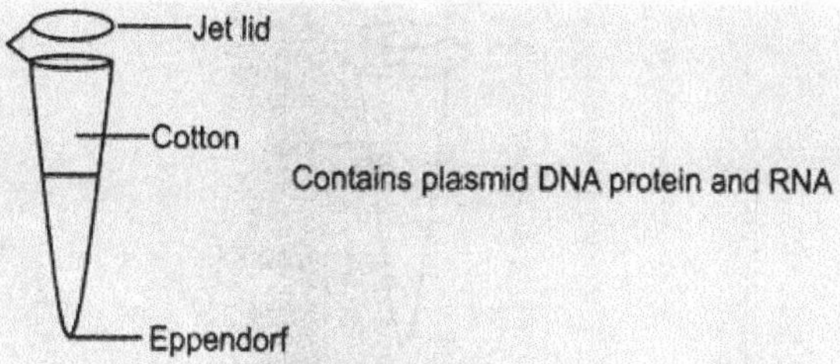

↓

Add isopropanol (cold) 300 µl in filtrate

↓

Mix well with 5 minutes. (after mixing solution must be appears clear, the presence of pellets represents chromosomal DNA.)

↓

Centrifuged at 5000 rpm for 10 min or 10,000 rpm for 5 min.

↓

Discard supernatants carefully

↓

Dry pellets by the use of tissue paper open eppendorf to evaporate isopropanol.
(Note: One eppendorf is kept and one eppendorf processed further)

↓

Add 200 µl of TES (Tris-EDTA-Sodium chloride buffer) and 1 µl RNase.

↓

Incubate at room temperature for 5 min.

↓

Add equal volume of phenol to remove protein.

↓

Mix gently for 5 min.

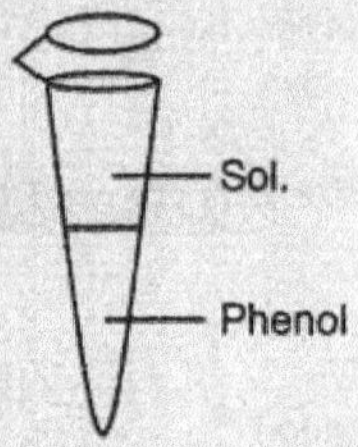

Centrifuge 3000 rpm for 5 min.

↓

Phenol chloroform step

Transfer solution to another eppendorf. It is most sensitive step.

↓

Add chloroform at equal volume to remove phenol.

↓

Mix gently for 5 min.

↓

Centrifuged 3000 rpm for 5 min.

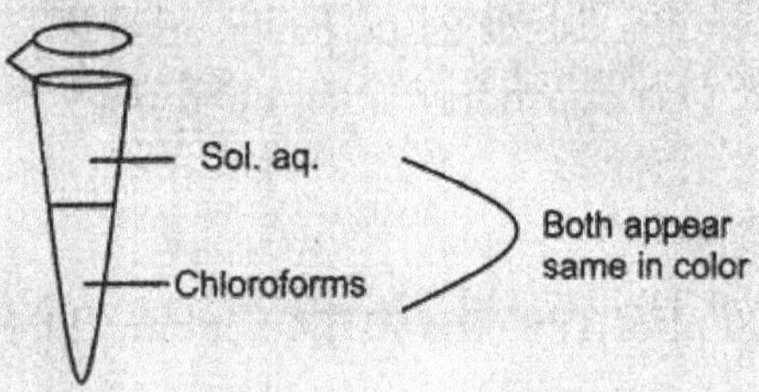

Alcohol ppt step

Transfer aqueous solution in other eppendorf. The step is sensitive and collect aquous solution most carefully.

↓

Add 1/10 vol. of 3 M sodium acetate (pH=5.1) mix and then add 2 vol. of ethanol.

↓

Keep eppendorf at –20°C overnight or in freezer.

Some people use ether to remove lipid between chloroform and alcohal treatments. But the step provides a point of precaution since ether present in upper layer and aquares solution containing plasmid DNA is present in lower region.

Some important facts for extraction of plasmid are as follows :

1. The penalty for failing to remove all traces of medium from the bacterial pellet in a preparation of plasmid DNA is resistant of cleavage by restriction enzymes. This is because of cell wall components in the medium, which inhibit the action of many restriction enzymes. This problem can be avoided by resuspending the bacterial pellet in ice cold STE (0.25 × volume of the original bacterial culture) and further centrifugation.
2. Some investigators find the extraction with phenol:Chloroform to be unnecessary. However, the elimination of this step sometimes causes problem as the extracted DNA is resistant to cleavage by restriction enzyme.
3. Lysozymes are a family of enzymes that catalyze the acid base hydrolysis of β (1–4) linkage between N-acelyl glucosamine and N-acelymuramic acid residues in the peptidoglycan of bacterial cell wall. After keeping eppendorf (10 μl) overnight the plasmid DNA is used for :

   (A) Electrophoresis (1% Agarose gel, 5 μl)

   (B) Cloning (5 μl)

Preparation of Solution

Solution I

- 50 mM glucose
- 25 mM Tris, pH8
- 10 mM EDTA

Solution II

- 0.2 M NaOH
- 1% Sodium dodecyl sulphate (SDS)
- Prepare 2M NaOH and 10% SDS stock and prepare required concentration fresh before use.

Solution III

3 M Potassium acetate brought to pH 5.5 with glacial acetic acid. Autoclave and store in a refrigerator.

TES

- 10 mM Tris-Cl, pH 8.0
- 100 mM NaCl
- 1 mM EDTA

Ether (Water saturated)

- 50 ml H_2O and 50 ml diethylether mix and store in cold.

3 M Sodium acetate (pH 5.2)

Dissolve 40.81 g of sodium acetate in 80 ml water adjust pH to 5.2 with glacial actic acid. Make up the volume to 100 ml. Autoclave and store.

Chapter 5

RESTRICTION ENDONUCLEASE DIGESTION

5.1 INTRODUCTION

Restriction endonuclease are bacterial enzymes that cleave specific nucleotide sequence in double stranded DNA with the production of defined fragments. Enzymes are named according to the bacteria from which they are first purified. In bacterium the enzymes prevent the infection of cell from foreign material enters by the process of transduction or transformation hence acts as interferon. If the same sequence of genetic material is present in its own sequence of bacteria the lysis is immunized by the methylation of adenine and cytosine.

Each enzyme has an optimal reaction buffer. The recommended reaction conditions are normally followed after the supplier's assay sheet. DNA should be pure and necessarily free from phenol, chloroform, salts, detergents and chelating agents.

The amount can be varied from 500 µg to 1 µg depending upon the aim of the experiment. BSA is routinely included in restriction digests to stabilize low protein concentrations and to protect against factors that cause denaturation.

Many enzymes make staggered cuts (with over hanging or sticky ends) that can base pair with DNA fragments generated by the same enzyme and form recombinant molecules. Other enzymes cut to yield blunt ended cleavage products, which can be ligated to other blunt ended fragments irrespective of restriction enzymes used.

5.2 METHODS FOR SELECTIVE CUT

The restriction endonucleases are divided into three classes. The enzymes belong to class-I and class-II have different recognition site and restriction site whereas enzymes belong to class-III have same recognition and restriction site. Each cutting site is characterized by the presence of palindrome sequence. The number of nucleotide of palindrome sequence is varied from 4–6 nucleotides. The cleaved segments of DNA may have staggered end or blunt end. The former is also known as protruding end or cohesive terminus or sticky terminus or overhanging end. The overhanging may has 5′ prime overhanging or 3′ prime overhanging.

5′ – GAATTC – 3′
3′ – CTTAAG – 5′
↓ EcoRI (*Escherechia coli* RI)
5′ – G + AATTC – 3′
3′ – CTTAA G – 5′
Creates 5′ prime overhanging

5′ – GGTACC – 3′
3′ – CCATGG – 5′

KpnI (Klebsiella pneumonia)
5′ – GGTAC + C – 3′
3′ – C CATGG – 5′
Creates 3′ prime overhanging

Sometimes two different restriction endonucleases have same recognition palindrome sequence but different cutting sites. They are known as iso-schizomers.

5′ – CCCGGG – 3′
3′ – GGGCCC – 5′

Sma I | Xma I

Sma I	Xma I
5′ – CCC + GGG – 3′	5′ – C + CCGGG – 3′
3′ – GGG + CCC – 5′	3′ – GGGCC + C – 5′

5′ – GTCGAC – 3′
3′ – CAGCTG – 5′

Sal I	Acc I	Hine II
5′ – G + TCGAC – 3′	5′ – GT + CGAC – 3′	5′ – GTC + GAC – 3′
3′ – C + AGCTG – 5′	3′ – CA + GCTG – 5′	3′ – CAG + CTG– 5′

Group of restriction enzymes operate under different conditions of salts and pH. For optimal activity, several different buffers are needed for the enzymes used in this process. To simplify procedures, we use a compromise restriction buffer that provides a compromise conditions preferred by various enzymes. All buffers are used at a final concentration of 1X. By use of the standard $C_1V_1 = C_2V_2$ formula to determine how much buffer is added to obtain a 1X solution. For convenience, use 2X buffer whenever possible. It can save a pipetting step to add water to bring a reaction up to 10 µl total volume. It is also easier and more accurate to pipette 5 µl than to pipette 1 µl. Compare a typical restriction reaction using 2X versus 10X restriction buffer.

Component	2X Buffer	10X Buffer
DNA	4 µl	4 µl
Enzyme	1 µl	1 µl
Buffer	5 µl	1 µl
Water	----	4 µl

The restriction enzymes are supplied as units in 50% glycerol.

500 unit/1000 µl

5000 unit/1000 µl

50000 unit/1000 µl

One unit of enzyme is the amount of enzyme necessary for the cutting of 1 µg of λ DNA within 1 hour at 37°C.

Procedure

1. Preparation of reaction mixture for DNA digest by a restriction endonuclease.

DNA (1 µg)	–	2 µl	
10X reaction buffer	–	2 µl	
10% BSA	–	2 µl	
Dist. H_2O	–	13 µl	simply add 1 µl of enzyme without diluting
Restriction enzyme	–	1 µl	even if it is excess.
Total	=	20 µl	

   Use three folds excess of restriction enzymes. Calculate the units required, based both on the number of restriction sites in the DNA and the size of the DNA.
2. Mix the digest by gently flicking the microcentrifuge tube. Do not vortex.
3. Pellet any droplets by spinning the tube for 3 seconds in a microcentrifuge at top speed.
4. Incubate the digests at 37°C in a water bath for 1–2 hour.
5. Stop the reaction by inactivating restriction enzyme by heating at 65°C for 10 min or adding 1 µl of 200 mM EDTA (pH 8.0).
6. Check the digestion by running in 0.8% agarose gel (Chapter-10).

Steps of Restriction Reaction

Take DNA/plasmid (1 µg) in an eppendorf (5 µl).

↓

Add 10X buffer (1 µl)

↓

Add distilled water (2 µl)

↓

Add RE 1U/µg - 2 µl

↓

Water is compensated to make total volume 10 μl after adding RE

↓

Momenting 37°C for 2hrs (varies for different RE check from catalog.)

↓

Stop enzyme action by heating at 65°C for 5 min.

It is important to determine the unit value of restriction enzyme available in vials and volume of restriction enzyme necessary for the cutting of DNA present in sample.

Unit determination of RE in vials

$$1000\ \mu l = 500\ U \text{ or } 1\ \mu l = 500/1000 = 0.5\ U$$

$$1000\ \mu l = 5000\ U \text{ or } 1\ \mu l = 5\ U$$

Determination of volumes for cutting

$$\left.\begin{array}{r} \text{DNA } (5\ \mu g) = 10\ \mu l \\ \text{10X buffer} = 5\ \mu l \\ \text{RE} = 25\ \mu l \\ \text{D water} = 10\ \mu l \end{array}\right\} 50\ \mu l$$

1% BSA (Bovine serum albumin) is added to provide stability to enzyme. Practically it is not compulsory.

After cutting the DNA is divided into two parts, one part is for verification of cutting and one part is for process of ligation.

5.3 RESTRICTION MAP

The breaking of DNA into fragments by the use of enzyme restriction endonuclease provides basis of the construction of restriction map. It is always constructed linear first and thereafter circular nature is determined. For the construction of map DNA is initially digested by single restriction enzyme and finally by double restriction enzymes. The priority in double digestion is determined by salt concentration, the low salt buffer containing restriction endonuclease enzyme is used first. When both RE enzymes have same buffer, we can use both enzymes simultaneously. For example, a 23 kb DNA is lysed by three RE enzymes separately and represent following fragments in gel electrophoresis.

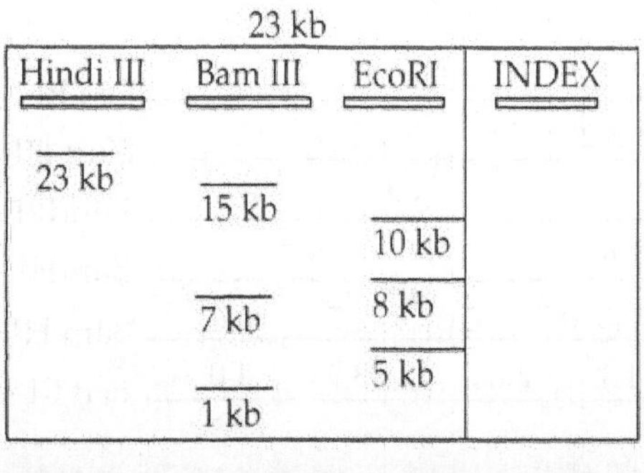

On the basis of above fragments restriction map is constructed as :

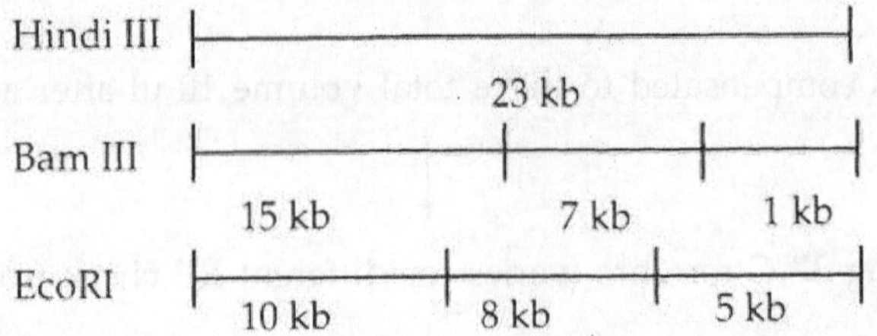

The construction of restriction map is further extended the fragmentation of double digestion by restriction endonuclease. For the construction of restriction map at first determine the volume of each reagent and add in the eppendorf. The volume of buffer and enzyme to be used are listed below. Calculate the volume of plasmid DNA that contains 1 µg. Calculate the volume of water required to keep the total volume of each reaction to 20 µl.

Components	Single Enzyme	Double Enzyme
DNA (1 µg)	---- µl	---- µl
10X reaction buffer	2 µl	2 µl
10% BSA	2 µl	2 µl
Dist. H_2O	--- µl	---- µl
Restriction enzyme1	1 µl	0.5
Restriction enzyme2	0	0.5 µl
TOTAL	20	20

Suppose a circular plasmid DNA molecule of size 10.5 Kb pair is studied and digested by restriction endonuclease Bam HI, EcoRI and HIND III singly and in all possible combinations. After lysis the restriction fragments are observed in electrophoresis as represented below.

Bam HI	Eco RI	HIND III	Bam HI+Eco RI	HINDIII+BamHI	Eco RI+HIND III
7.3	10.5	5.1	6.7	4.6	4.0
3.2		3.4	3.2	2.7	3.4
		2.0	0.6	2.0	2.0
				0.7	1.1
				0.5	
10.5	10.5	10.5	10.5	10.5	10.5

On the basis of above data at first linear line fir each is constructed and thereafter circular nature is determined.

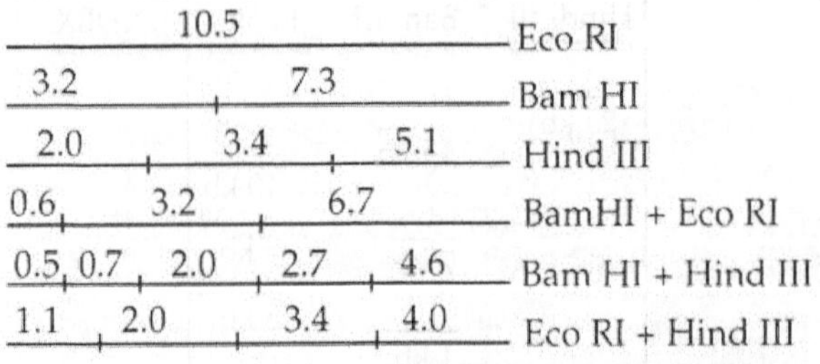

In the construction of restriction map for the plasmid the letter 'B', 'E' and 'H' are used for enzymes Bam HI, Eco RI and HIND III respectively. The map is now constructed as follows:

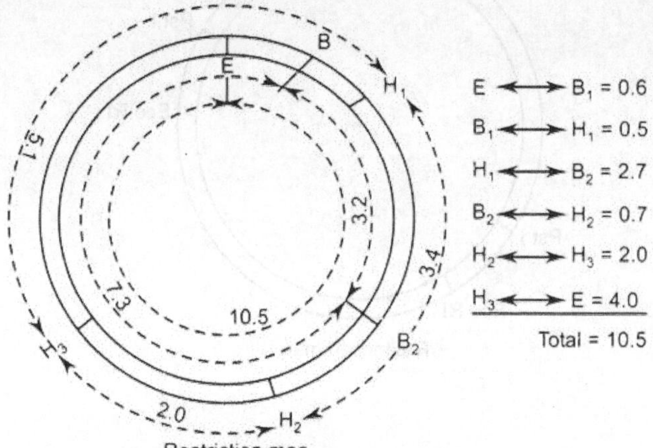

Restriction map

In next condition, suppose a circular plasmid DNA molecule of size 12 Kb pair has been studied and digested by restriction endonucleases Pst I, EcoRI and Xma I in single and double digestion. After lysis the restriction fragments are analyzed by gel electrophoresis and observed as follows :

Pst I	Eco RI	Xma I	Pst I+Eco RI	Pst I +Xma I	Eco RI+Xma I
6.0	4.0	12	5.0	6.0	7.0
			3.0	5.0	3.0
			1.0	1.0	2.0
12	12	12	12	12	12

Here at first we calculate the number of fragments-

$$\text{Pst I} = 12/6 = 2 \text{ and total} = 6 + 6 = 12$$

$$\text{EcoRI} = 12/4 = 3 \text{ and total} = 4 + 4 + 4 = 12$$

$$\text{Pst I} + \text{EcoRI} = 5 + (3 \times 2) + 1 = 12$$

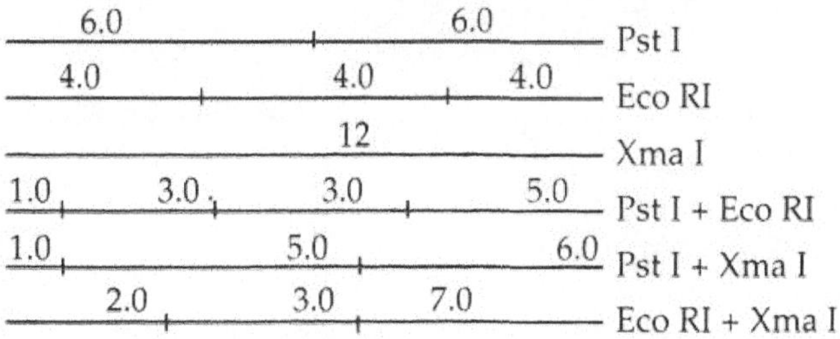

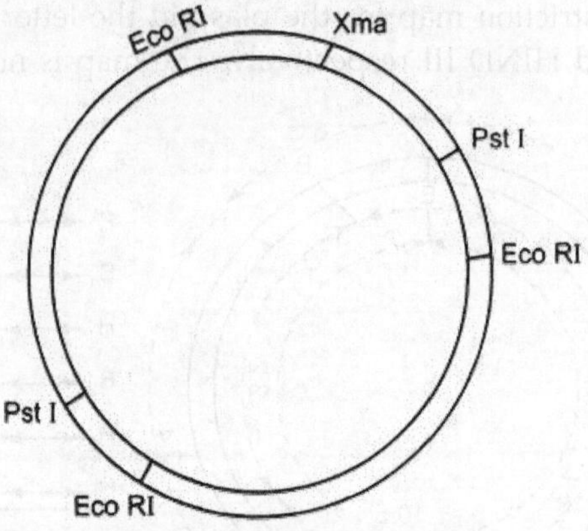

Restriction map

Chapter 6

LIGATION

6.1 INTRODUCTION

Some restriction endonuclease enzymes cleave asymmetrically within the recognition sequence of plasmid DNA and target DNA. When the protruding sequences are compatible, they can anneal to form a linear hybrid molecule whose two parts are held together by pairing between the bases in the protruding termini. Formation of a circular recombinant plasmid capable of transforming *E.coli*, therefore, occurs in a two steps relation.

1. An inter-molecular relation between linear plasmid and incoming DNA, which generates a non-covalently bonded, linear hybrid.
2. An intra-molecular reaction, in which the protruding termini of the linear hybrid are joined together, forming a non-covalently bonded, circular recombinant molecule.

Annealing brings the 5′-phosphate and 3′-hydroxyl residue on vector and target DNAs into close alignment, which allows DNA ligase to catalyze the formation of phosphodiester bond.

Ligation reactions should be designed with care so as to maximize the yield of circular monomeric recombinants. The first, intermolecular stage of reaction requires high concentration of DNA, whereas the second intra-molecular reaction works most efficiently when the concentration of DNA is low. However as a general rule, acceptable yield of monomeric circular recombinants can usually be obtained from ligation reactions containing equimolar amounts of plasmid and target DNA, with the total DNA concentration <10 µg/ml. If the molar ratio of plasmid vector to target DNA is incorrect, then the ligation reaction may generate an undesirably high proportion of empty plasmids carrying tandem inserts of foreign DNA.

Sometimes yield of circular monomeric recombinants is increased by the use of two restrictions enzyme instead of one. In this case foreign DNA fragment is produced by digestion with two restriction enzymes with different recognition sequence. In this case the termini of foreign DNA fragment will be non-complementary and unable to ligate to each other. The segment of DNA quickly ligate to plasmid digested be any one of two restriction enzymes, generating a high yield of circular recombinants containing a single insert in a predefined orientation. This process is referred as forced ligation.

6.2 METHOD FOR JOINING OF DNA

T_4 DNA ligase catalyze the synthesis of a phosphodiester bond between adjacent phosphotermini in two different ends of the same or two different DNA molecules. It requires Mg^{2+} and ATP as cofactors. Double stranded DNA molecules with complementary cohesive ends or blunt ends with 3'-OH and 5'–P termini can be used as substrate by T_4 DNA ligase. The different methods of ligation are available as mentioned below but in common laboratory we use ligation by end filling.

1. Ligation by end filling.
2. Ligation by the use of linker molecules. It is commonly used for PCR products.
3. Ligation by use of adaptors. They are per cut linkers commercially available in market.
4. Ligation by homopolymer tailing :

Ligation Reaction

For the ligation we simply take genomic componants and necessary chemicals in an eppendorf and incubate.

Digested Vector DNA (20 ng)	–	1 µl
Insert DNA (3 times excess)	–	1 µl
10 X ligation buffer	–	1 µl
10 mM ATP	–	1 µl
T_4 DNA ligase (15 units)	–	1 µl
Water	–	5 µl
Total	–	10 µl

Incubate at 10 to 12°C for 4 to 8 hours or at 16°C for 24 hours.

The volume of vector DNA, insert DNA and water could be changed without altering the total volume. Two types of ligase are commercially available.

T_4 DNA ligase

E.coli ligase

Steps for Ligation Reaction

Ligase enzyme is supplied in glycerol Ligase Reaction (20 µl)

Vector/plasmid DNA	=	× vol
Gene	=	× vol
10 X Ligase buffer	=	2 µl
10 nM ATP	=	1 µl
Ligase	=	1 µl
H_2O	=	....

↓

Mix and keep in 12–20°C over night or till we use.

During ligation reaction more gene is required with respect to vector, otherwise the condition can caused self-ligation. Chemicals are provided by GeNi company and Biolab. The volume of vector and gene is determined quantitatively. Always maintains excess of desired gene. For example, if the ligation reaction is of 30 µl, we calculate ligating chemicals as follows:

Ligation reaction (30 µl)

Hind III cut plasmid DNA	=	2 µl
Gene (λ DNA)	=	0.2 µl
10 mM ATP	=	1 µl
Enzyme	=	1 µl
10 X ligase buffer	=	3 µl
H_2O	=	4.8 µl

↓

Take in eppendorf

↓

Mix well and keep overnight

A. If size of vector is 2 kb and gene is 0.5 kb

Commonly 1 ml contains 1 µg DNA. But when 1ml contains 500 ng the gene used is 125 ng.

i.e., 0.5/2 = 125

B. If size of vector is 1 kb and gene is of 2 kb

Gene = 200 ng Vector = 100 ng.

The identical sticky ends of DNA fragments can provides following conditions during ligation reaction.

1. Plasmid DNA can self-ligate without an insert.
2. Chromosomal DNA fragments can self-ligate or can ligate each other.
3. A plasmid DNA ligate with one or more chromosomal fragments.

Chapter 7

Competent Cell and Transformation

7.1 INTRODUCTION

The recombinant DNAs do not enter into bacteria by their own power, but require assistance traversing the outer and inner cell membranes and reaching an intracellular site where they can be expressed and replicate. The methods that have been devised to achieve these goals fall into two classes chemical and physical.

E.coli cells washed in cocktails of simple salt solution achieve a state of competence during which DNA molecules may be admitted to the cell. Most of the chemical methods currently used for bacterial transformation are based on the observations of *Mandel* and *Higa* (1970), who showed that bacteria treated with ice cold solutions of calcium chloride and then briefly heated to 37°C or 42°C could be transfected with bacteriophage λ DNA. The same method was subsequently used to transform bacteria with plasmid DNA (*Cohen et al* 1972) and *E.coli* chromosomal DNA (*Oishi* and *Cosloy* 1972). From these works a number of literatures are available for preparation of competent cells. The variations include using complex cocktails of divalent cations in different buffers, treating cells with reducing agents, adjusting the ingredients of the cocktail to the genetic constitution of particular strain of *E.coli*, harvesting cells at specific stage of the growth cycle, altering the temperature of the culture before exposure to chemicals, optimizing the extent and temperature of heat shock, freezing and thawing of cells and exposing cells to organic solvents after washing in divalent cations. By all these treatments, it is now possible on a routine basis to achieve transformation frequencies ranging from 10^6 and 10^9 transformants/µg of superhelical plasmid DNA. However, the improvements in transformation frequency are a tribute to the power of empirical experimentation. How these combinations of chemical agents and physical treatments induce a state of competence remains as obscure.

There are two ways to obtain stock of chemically induced competent *E.coli*. The first option is to purchase frozen competent bacteria from a commercial source and the second is to develop competent cell in laboratory by using of standard strains of *E.coli* (Section 15.4).

7.2 COMPETENT CELL PREPARATION

1. Start an overnight culture of *E.coli* by using an inoculation loop to scrap from a single colony into 2 ml LB.
2. Do not add antibiotic to LB for competent cell preparation.
3. Place at 37°C in shaking incubator.
4. Add 0.1% to 1% from this preculture to the main culture flask with LB.
5. Grow for 2–3 hours. This culture at mid log can be used for competent cell preparation and plasmid isolation.

LB Medium (200 ml)

- Tryptone – 2 gms
- Yeast extract – 1 gm
- NaCl – 2 gms

Take Tryptone , Yeast extract, NaCl and mix it in 120 ml of double distilled water then adjust the pH with 2 m NaOH as 7.5.

LB Agar (100 ml)

Take 100 ml of LB and add 2% agar *i.e.,* 2 gms and autoclave. Concentration of antibiotics to be used:

- Ampicillin – 100 μg/ml
- Kanamycin – 50 μg/ml
- Tetracycline – 10 μg/ml

Aseptic Culture

1. Incubator
2. *Laminar flow Hood:* The laminar flow hoods are of two types namely horizontal for non-pathogenic bacteria and vertical for pathogenic bacteria. In later HEPA (highly efficient particulate air) filter is used . For highly pathogenic microbes, gloved chamber is used. UV light is used for the sterilization of laminar flow chamber. The chemical used for the sterilization of room/chamber is ammonium salts placing in different places in room. It is advisable to switch on UV light for 45 min. before entry and switch off just before the entry into the room.

   The thermolabile chemicals are sterilized by filter sterilization. There are multi-pore filters of 0.22 μ and 0.45 μ. But nowaday disposable plastic filters are available in market for 250 ml. However, it is not useful for sterilization of medium for animal cells. In the absence of autoclave, the thermal sterilization is also completed by the use of pressure coocker but it is advisable to maintain pressure by adjusting flame of fire. During sterilization the scaling of test tube or flask is also crucial.

Procedure

1. Take *E.coli* cells of mid log phase which is in the range of 0.4 to 0.5 OD at 600 nm.
2. Centrifuge the cells for 5 min. at 5000 rpm, discard the supernatent and take the pellet.
3. To the pellet add pre-cooled 50 mM $CaCl_2$ solution (equal culture vol.) and suspend the cells very gently.
4. Centrifige at 5000 rpm, 4°C for 5 min.
5. Resuspend the cells with pre-cooled 50 mM $CaCl_2$ (1/10 th culture vol.) and leave it in ice. The cells are now competent for taking up exogenous DNA.

7.3 TRANSFORMATION

The common method of bacterial transformation utilizes high levels of calcium chloride to facilitate the entry of plasmid DNA vectors into *E.coli* in conjuction with structural alternation of the bacterial cell wall. In general, there are three basic steps in the introduction of plasmid DNA into cells—(1) Preparation or 'Transformation of these competent' cells to enable cells to accept DNA, (2) Transformation of these competent cells, and (3) Selection of transformants. Most *E.coli* strains exihibit transformation efficiencies (measured as transformants per micro gram plasmid DNA) in the range of 10^5 to 10^8. The factors influencing this efficiency depends upon the condition that render cells competent.

(1) These are harvesting cells during the logarithmic phase of growth.

(2) Maintaining cells at a temperature of 4°C during treatment.

(3) Prolonged exposure to ice cold $CaCl_2$.

Common concept for transformation is based on four major steps. The steps are as follows:

1. *Preincubation:* Cells are suspended in a solution of cations and incubated at 0°C. The cations are thought to complex with exposed phosphates of lipids in the *E.coli* cell membrane. The low temperature freezs the cellmembrane, stabilizing the distribution of changed phosphates.
2. *Incubation:* DNA is added and the cell suspension is further incubated at 0°C. The cations are thought to neutralized negatively charged phosphates in the DNA and cell membrane. The DNA molecule is free to pass through the cell membrane due to the charge neutralization.
3. *Heat shock:* The cell-DNA suspension is briefly incubated at 42°C and then returned to 0°C. The rapid temperature change creates a thermal imbalance on either side of the *E.coli* membrane which is thought to create a drift that sweeps plasmids into the cell.
4. *Recovery:* LB broth is added to the DNA-cell suspension and incubated at 37°C, ideally with shaking, prior to plating on selective media. Transformed cells recover from the treatment, amplify the transformed plasmid and begin to express the antibiotic resistance protein.

The incubation and heat shock steps are crucial. Since preincubation and recovery steps do not consistently improve the efficiency of colony transformation, they have been non-applicable in the most of the laboratory. If time permits, a preincubation of 5–15 minutes and/or a recovery of 5–30 minutes may be included. Transformation vary between strains of *E.coli* and the vectors.

Factors such as plasmid size, quantity, purity, DNA configuration and selectable marker also influence the out come of a transformation experiment.

Procedure

1. Take 10 µl ligated recombinant DNA.
2. 100 µl of 100 mM $CaCl_2$ in an eppendorf tube.
3. Add 100 µl of competent cells.
4. Leave in ice fo 30 min.
5. Give a heat shock at 42°C for 1 min.
6. Add to 2 ml of plain LB in 30 ml Screweap tube.
7. Keep inside the Incubator for incubation. Incubate for 1 hr at 37°C.
8. Plate 100 µl of the expressed culture in Amp., X-Gal, IPTG plate.

The following reaction are used for transformation:

1. Competent cells alone in LB to check survival of competent cells.
2. Competent cells with uncut vector in LB + antibiotic + X-gal + IPTG to check the competence of *E.coli*.
3. Competent cells with cut vector in LB + antibiotic + X-gal + IPTG to check whether digestion is complete.
4. Competent cells with self-ligated vector in LB + antibiotic + X-gal + IPTG to check whether ligation reaction is efficient.
5. Competent cells with cut vector and insert ligated in LB + antibiotic + X-gal + IPTG to identify recombinant clone. X-gal (5-bromo-4-chloro-3-indolyl-β-D-galatopyranoside) : Prepare a stock in dimethyl formamide.

Working concentration = 40 µg/ml

IPTG (Isopropylthiogalactoside) :

IPTG should be used as an inducer of lac operon if *E.coli* cells, such as JM 103 and 109, are used for transformation.

Prepare a stock in sterile distilled water.

Working concentration = 96 µg/ml

Steps for Transformation Procedure

Host is always plasmid DNA (available commercially)

↓

Grown in LB medium for 2½ hr. (OD_{600} = 1 OD)

↓

To take 1 ml culture in eppendorf in sterilyzed condition.

↓

Centrifused at 3000 rpm for 5 min.

Discard supernatent

↓

Add 100 µl of 50 mM $CaCl_2$ and incubate 5 min in ice

↓

Suspend gently

↓

Centrifused 3000 rpm for 5 min (ice)

↓

Discard supernatent and take pellet

↓

Add 50 mM $CaCl_2$ (100 µl) (keep in ice)

↓

Now competnet cells are formed.

↓

Add 100 µl of 100 mM $CaCl_2$ and mix with *r*-DNA (10 µl)

↓

Both take in eppendorf and Incubate for 5 min. in ice

↓

Give heat shock at 42°C

↓

Incubate 5 min. in ice

↓

Add 2 ml LB in drops

↓

Incubate 1 hour { ½ h without shaking; ½ h with shaking }

↓

Plate 5 µl in LB agar

↓

Incubate overnight and observe colonies

After the transformation and sreak plating (Section 15.2) the transformation efficiency is determined.The transformation efficiency is expressed as the number of antibiotic resistant colonies per microgram of plasmid DNA. The different componants for determination of efficiency are as follows:

$$\text{Total mass of plasmid } (\mu g) \text{ (TMP)} = \text{concentration} \times \text{Volume}$$

$$\text{Fraction spread (FS)} = \frac{\text{Volume supension spread}}{\text{Total volume suspension}}$$

$$\text{Mass plasmid spread (MPS)} = \frac{\text{Total mass plasmid}}{\text{Fraction spread}}$$

$$\text{Transformation efficiency (TE)} = \frac{\text{Colonies observed}}{\text{Mass plasmid spread}}$$

Chapter 8

SCREENING OF CULTURE

8.1 INTRODUCTION

The screening is a laboratory method for the analysis that whether the transformed bacteria carry a recombinant plasmid or an empty vector to determine transformation efficiency. The method is based on the nature of marker available in vector plasmid or marker site and its secondary metabolic activity. Plasmid cloning vector generally contains at least one antibiotic resistant (Chapter-4) gene that can be used to select bacterial cells that have taken up the recombinant plasmid. Antibiotic resistant genes code for proteins that either inactivate the antibiotic or prevent entry of the antibiotic into the bacterial cell. The specific feature provides basis of screening of bacterial cells after transformation. The two methods are used in laboratory to identify the transformed bacteria.

8.2 ANTIBIOTIC SELECTION

Screening is based on the principle of insertional inactivation. For example, plasmid pBR 322 has resistant site for ampicilin and tetracycline. The ampicilin resistant site is lysed by restriction endonuclease Pst I and tetracycline resistant site is lysed by restriction endonucleases Hind III and Bam HI. Single site within the any one of antibiotic resistance marker could be used for cloning. For example,

Pst I – site within the Amp^r marker

Hind III/Bam HI – within the Tet^r marker

Selection can be occurred by patching the cells in both antibiotic plates. If a gene is inserted in the Pst I site, the transformants become amp sensitive (Amp^s) and will not grow in an Amp^r Plate. But the corresponding patch will grow in tet^+ plate. By comparing both the plates, the parent and the recombinant clones can be differentiated. Tetracycline enters bacterial cells by passive diffusion across the outer membrane through porin channels, which are composed of the Omp F protein.The chief mechanism by which *E.coli* becomes resistant to high concentration of tetracycline involves antiporter proteins, known as tet-protein which is located in the bacterial inner membrane and, in exchange for a protein, expel intracellular tetracycline-metal complexes against a concentration gradient. Tet gene of pBR 322 and many other vectors encodes a 392 residue polypeptide composed of two domains each containing six transmembrane segments.

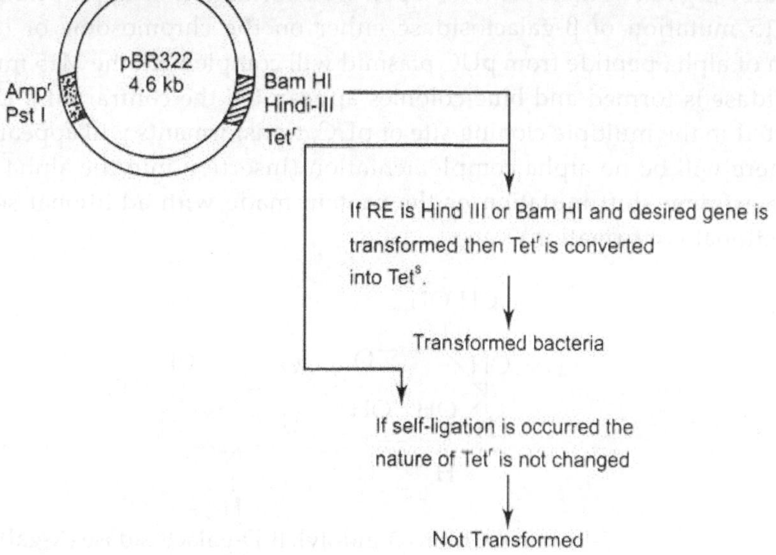

8.3 BLUE-WHITE SELECTION

Many plasmids also contain a screenable marker allowing visualization of the presence of insert DNA. Portions of the Lac operon have been used to create a visual screen for the presence of insert DNA in plasmids. The Lac operon contains a single promoter, an operator that binds repressor and regulates promoter function, and 3 genes encoding enzymes used for lactose metabolism. The last gene Lac Z encodes β-galactosidase, which cleaves lactose.

The pUC vectors replicate to high numbers , are small (2.7 kb) and allow blue-white screening for recombinant plasmids. They are ideal for propagating DNA smaller than about 10–15 kb. The pUC family of plasmid vectors offers a multiple cloning site (MCS) and the possibility of selecting chimeric plasmids with a coloration test based on functional complementation assays for β-galactosidase. Bacteria that carry the regulatory elements and the Z-gene of the Lac operon can be induced by isopropyl thiogalactoside (IPTG) to express β-galactosidase. This enzyme hydrolyzes the colorless 5-bromo-4-chloro-3-indoyl β-D-galactopyranoside (X-gal) and give rise to a blue insoluble indigo derivative. This enzymatic activity removes amino acid residues 11 to 41. This constitutes a portion of alpha peptide, which is defined as the region spanning amino acid residues 3 to 92 of β-galactosidase. It should be noted that alpha peptide is not a sub-units of the β-galactosidase but is only a portion of it. It has been shown that expression of alpha peptide in *E.coli* having M15 deletion (deletion of 11 to 41 amino acid led to complementation and functional β-galactosidase was found in these cells. This unique complementation is referred to as intracistronic complementation in which two non-functional polypeptides alpha peptide lacks the C terminal and M15 has a deletion at the N-terminal end) complement each others defect. Since the alpha peptide complements the deletion of M15 this a referred to as alpha complementation. The alpha complementation, that leads to the generation of a functional β-galactosidase, can be visualized by the appearance of blue colonies on indicator plates containing X-gal and IPTG in certain *E.coli* strains.

A whole series of plasmid vectors containing the bacterial Lac gene with multiple cloning site inserted at a position corresponding to the fifth amino acid of the alpha peptide. DNA fragments may be inserted in the unique sites of the MCS. Successful cloning is monitored by the loss of β-galactosidase activity upon transformation of appropriate host strains that have a M15 mutation of β-galactosidase either on the chromosome or on the F episome. Expression of alpha peptide from pUC plasmid will complement the M15 mutation, functional β-galactosidase is formed and blue colonies appear. On the contrary if a DNA fragment has been inserted in the multiple cloning site of pUC, transformants will appear as white colonies because there will be no alpha complementation (Insertion into the alpha peptide sequence may cause a frame shift mutation or the protein made with additional sequences may not attain functional confirmation).

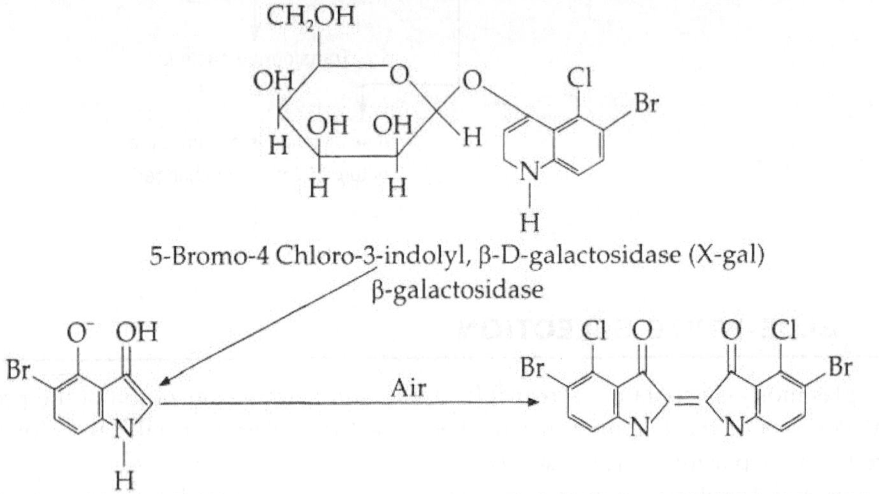

5-Bromo-4 Chloro-3-indolyl, β-D-galactosidase (X-gal)

5-bromo-4-Chloroindoxyl

5,5′-dibromo-4,4′-dichloroindigo

The recombinant plasmids could be distinguished from the parental plasmids by plating *E.coli* harboring them on a chromagenic compound called X-gal (5-bromo-4-chloro-3-indoyl-β-D-galactopyranoside). X-gal when cleaved by β-galatosidase, gets converted into a blue, insoluble precipitate, which makes the colonies to appear blue. Parental plasmids make functional β-galactosidase. Colonies of *E.coli* harboring the recombinant plasmids do not cleave X-gal since they lack β-galactosidase. Therefore, *E.coli* with recombinant plasmids appears white in color.

Material and Instrumentation

X-gal solution (2%W/V), IPTG solution (20%W/V), LB agar plate containing appropriate antibiotic, LB top agar, inoculating loop, transformed *E.coli* culture and isopropyl β-D-thyogalactoside. The isopropyl β-D-thyogalactoside is a non-fermentable analog of lactose that inactivates the lac Z repressor and therefore induce transcription of the lac operon.

Procedure

Add 40 µl of 2% X-gal solution and 7% µl of 20% IPTG solution (not compulsory) onto the center of a pre-made 90 mm agar plate (LB) containing the appropriate antibiotic. For a 15 cm diameter agar plate, transfer 100 µl of X-gal and 20 µl of IPTG to the center of the plate.

↓

Spread the solution over the entire surface of the plate by a sterile spreader.

↓

Inoculate the plate at 37°C until all the fluid has disappeared.

↓

Use inoculation loop and do streak plate culture or culture bacteria by spreading upto 100 µl bacterial suspension (90 mm plate)/200 µl bacterial suspension (15 cm plate)

↓

Incubate the plate in an inverted position for 12–19 hours at 37°C.

↓

Remove the plate from the incubator and store it at 4°C for several hours during which the blue color develops to its full extent. It represents self-ligation. The white color is appeared due to absence of β-gal gene represents both recombinant nature and non-availability of plasmid.

Chapter 9

HYBRIDIZATION

9.1 INTRODUCTION

Southern transfer and hybridization (Southern, 1978) is used to study how genes are organized within genomes by mapping restriction sites in segments of genomic DNA for which specific probes are available. Genomic DNA is first digested with one or more restriction enzymes, and the resulting fragments are separated according to size by electrophoresis through a standard agarose gel. The DNA is then denatured in sites and transformed from the gel to a solid support (usually a nylon or nitrocellulose membrane). The DNA attached to the membrane is hybridized to a labelled DNA, RNA or oligonucleotide probe and bands complementary to the probe. Thereafter, they are located by an appropriate detection system. By estimating the size and number of the bands generated after digestion of the genomic DNA with different restriction enzymes, singly or in combination, it is possible to place the target DNA with a context of restriction sites.

For almost twenty years, the only available support for immobilization of DNA was nitrocellulose, which was first used in powder form and later as sheets. Northern hybridization was initially carried out exclusively with RNA immobilized on activated cellulose papers. However, it was realized that RNA denatured by glyxoal, formaldehyde or methylmercuric hydroxide binds tightly to nitrocellulose. For several years, nitrocellulose therefore support of choice for both northern and southern hybridization. Despite its evident success, nitrocellulose is not an ideal meterial for solid phase hybridization.

9.2 METHOD FOR SOUTHERN HYBRIDIZATION

To identify which transformants contain the insert DNA, a purified DNA encoding the gene of interest is labelled and used to detect those clones that contain similar sequences. In a typical experiment of screening, a successful rate of 1 positive in 10,000 is reported.

After DNA has been transferred to nitrocellulose or nylon, detection of the DNA is carried out by hybridization to a probe DNA that is labelled with radioactive element or derivatized with a detectable group.

Hybridization and washes are performed at 10–15°C or below the temperature. For homologous DNA, 65°C is a common hybridization temperature. It is done under conditions that favor DNA binding, with 6X SSC overnight. Following hybridization the blots are washed under stringent conditions that destabilize interactions between DNA duplexes at 2X SSC and then at 0.2X SSC. The lower strength of the wash buffer lowers the melting temperature of a given DNA hybrid so that non-specific binding is eliminated. A southern blot involves DNA transfer to nitrocellulose and probed with DNA or RNA.

Procedure

1. Run a agarose/acrylamide gel with control and DNA digested with different restriction enzymes. Include an uncut DNA preparation, which is used for restriction digestion.
2. Run in 1X TEB at 60 V. Stain the gel in ethidium bromide and take a photo.
3. *Denaturation*: Soak the gel in 1 litre of 1 M NaCl + 0.5 M NaOH in a tray and shake for 50 min. at room temperature.
4. Wash 4 times with sterile double distilled H_2O (10 min each time).
5. *Neutralization*: Shake the gel in 1 litre of 1.5 M NaCl + 0.5 M Tris (pH = 7) shake for 45 min. at room temperature.
6. *Transfer*: Do capillary transfer with (positively charged) Nylon membrane or nitro cellulose membrane.
7. Remove blot with DNA side up and rinse one time with 2X SSC for 30 seconds.
8. After rinsing air dry the blot over a crude filter paper for 10–15 min.
9. Fix the DNA by exposing to UV for 3 min. for nylon membrane. Or bake in a vacuum oven at 80°C for 30 min. for nitrocellulose membrane.

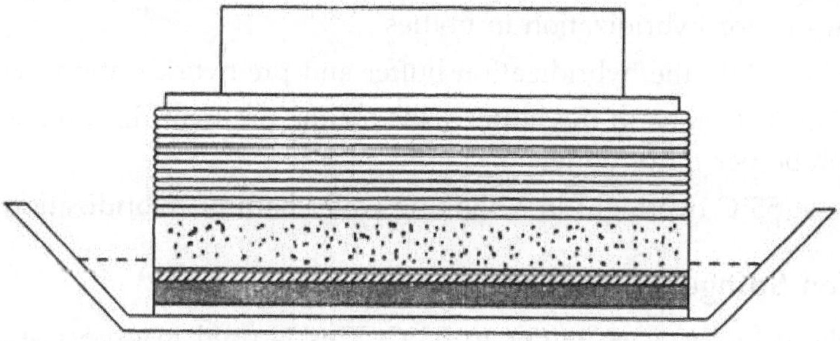

FIGURE 9.1. *Southern Blotting Apparatus (For blotting components are arranged as glass plate, agarose gel containing DNA, nitrocellulose membrane, 3 mm thick watmann no-3 paper, paper towels, glass plate and weight-from lower to upper direction)*

Labelling and Detection of DNA

Direct labelling of DNA with alkaline phosphatase and detection method using the Gene Images Alkphos Direct Labelling and detection system from Amersham Pharmacia Biotech Kit is followed here.

It involves directly labelling of probe DNA or RNA with a specially developed thermostable alkaline phosphatase enzyme. This is achieved by completely denaturing the probe so that it is in single-stranded form. The addition of the cross-linker covalently couples the enzyme to the nucleic acid probe. High target blots probed with the Alkphos Direct system can be detected with NBT/BCIP color detection reagent or by chemiluminescent detection or chemifluorescent detection.

Preparation of Labelled Probe

1. Dilute 20 µl cross-linker solution with 80 µl of the water supplied to give the working concentration.
2. Dilute DNA (or RNA) to be labelled to a concentration of 10 ng/µl using the water supplied.
3. Place 10 µl of the diluted DNA sample in a micro-centrifuge and denature by heating for 5 minutes in a vigorously boiling water bath.
4. Immediately cool the DNA in ice for 5 minutes. Spin briefly in a micro-centrifuge to collect the contents at the bottom of the tube.
5. Add 10 µl of reaction buffer to the cooled DNA. Mix thoroughly but gently.
6. Add 2 µl labelling reagent. Mix thoroughly but gently.
7. Add 10 µl of the cross-linker working solution. Mix thoroughly. Spin briefly in a micro-centrifuge to collect the contents at the bottom of the tube.
8. Incubate the reaction for 30 minutes at 37°C.
9. The probe can be used immediately or kept on ice for up to 2 hrs for long storage, labelled probes may be stored in 50% (v/v) glycerol at –15°C to –30°C for up to six months.

Hybridization

1. Pre-heat the required volume of prepared Alkphos Direct hybridization buffer to 55°C. The volume of buffer should be equivalent to 0.25 ml/cm^2 for large blots hybridized in plastic bags or for hybridization in bottles.
2. Place the blots into the hybridization buffer and pre-hybridization oven.
3. Add the labelled probe to the buffer used for the pre-hybridization step. Typically use 5–10 ng probe per ml of buffer.
4. Hybridize at 55°C overnight in a shaking water bath or hybridization oven.

Post Hybridization Stringency Washes

1. Preheat the primary wash buffer to 55°C. This is used in excess at a volume of 2–5 ml/cm^2 of membrane.
2. Carefully transfer the blots to this solution and wash for 10 minutes at 55°C, with gentle agitation.
3. Perform a further wash in fresh, primary wash at 55°C for 10 minutes.
4. Place the blots in a clean container and add an excess of secondary wash buffer. Wash, with gentle agitation, for 5 minutes at room temperature.
5. Perform a further wash in fresh, secondary wash buffer at room temperature for 5 minutes.

Color Detection

1. Add 45 µl NBT stock and 35 µl BCIP stock to 10 ml color detection buffer.
2. Drain off any excess wash buffer from the blot (by touching the corner of the blot against the box used for washing) and place it (sample side up) on to a non-absorbant surface such as saran wrap.

3. Pipette the detection reagent mix on to the blots (0.125 ml/cm^2 of membrane) and incubation will probably be required for optimum sensitivity.
4. Once the required color development has been achieved, rinse the blots for 2 × 2 minutes in distilled water and leave on an absorbent surface to dry.
5. It may be desirable to photograph the results as the color will fade over time.

Color Detection Buffer

- 100 mM Tris–HCL
- 100 mM NaCl
- 50 mM $MgCl_2$
- Adjust pH to 9.5

9.3 WESTERN HYBRIDIZATION

Identification of a specific protein in a complex mixture of proteins or antibodies can be accomplished by western blot. In this technique protein is electrophoretically separated on a polyacrylamide slab gel. The protein bands are transferred into a nitrocellulose membrane by electrophoresis or by diffusion such that the membrane gets a replica of the gel containing protein bands. The individual protein bands are indentified by flooding the membrane with primary antibody. After this the antigen-antibody can be detected by adding a secondary antibody conjugated with enzyme. The band is visualized by adding substrate which being a chromogen or conversion to product gives color. The intensity of color shows the quantity of antigen-antibody complex formed.

Enzymes and Substrate for Western Blot Analysis

A number of enzymes have been employed for western blot developing-alkaline phosphatase and horse radish peroxidase.

Alkaline Phosphotase

Solution A

- NBT-Nitroblue tetrazolium
- 3 mg in 0.25 ml buffer containing 0.1 M Tris.HCl pH 8.8, 0.1 M NaCl, 0.005 M $MgCl_2$.

Solution B

- BCIP-(5-bromo-4-chloro-indolyl phosphate)
- 1.5 mg in 10 μl dimethyl sulphoxide, add 0.25 ml of 2 M Tris. HCl pH 9.8.
- Mix A and B and makeup to 10 ml.

Horse radish peroxidase

- 30 mg of 4-chloro-1-naphthol in 10 ml methanol
- 30 ul of H_2O_2 in 40 ml TBS

Mix A and B at room temperature.

Transfer of Proteins by Diffusion

Transfer of protein bands from an acrylamide gel onto a membrane is occurred by diffusion. The transfer by the capillary action of buffer is known as capillary blotting.

Solution to be Prepared

1. *Urea buffer*
   - Sodium chloride = 0.29 g
   - EDTA = 0.07 g
   - Tris = 0.12 g
   - Urea = 24.00 g

   Adjust the pH to 7.0 and make up the volume to 100 ml with distilled water.
2. *Transfer buffer*
   - Sodium chloride = 2.9 g
   - EDTA = 0.7 g
   - Tris = 1.2 g

   Adjust the pH to 7.0 and make up the volume to one litre with distilled water.
3. *Phosphate buffered saline (FBS) pH 7.2*
   - Sodium chloride = 8.00 g
   - Potassium chloride = 0.20 g
   - Disodium hydrogen = 1.15 g
   - Phosphate
   - Potassium dihydrogen = 0.20 g
   - Phosphate
   - Distilled water = 1 litre.

Procedure

1. Remove SDS from the gel by washing the gel with urea buffer for 30 min.
2. Measure the size of the gel and cut Watmann no. 3 paper 6 sheets and nitrocellulose membrane 2 sheets.
3. Assemble the sandwich as given in figure.
4. Put rubber bands to keep the sandwich intact.
5. Immerse the whole sandwich in transfer buffer kept in a tray.
6. Keep a light weight on top to prevent the floating of the sandwich and to make sure that the sandwich is kept immersed completely. Avoid traping any air bubble while preparing the sandwich assembly.
7. Keep for transfer for 12 to 18 hrs.
8. Remove nitrocellulose membrane and proceed for antibody labelling or can be stored in a box for more than 6 months.

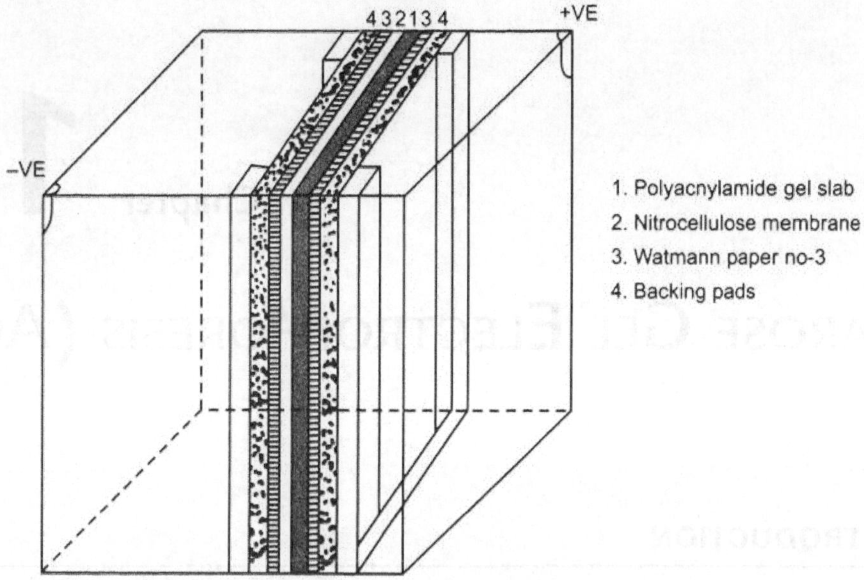

FIGURE 9.2. *Electroblotting Apparatus (The western blotting sandwich is enclosed in a plastic frame containing backing pads, 3 mm thick gel, nitrocellulose membrane 3 mm thick watmann paper and backing pad)*

Electrophoretic Transfer

1. Soak the gel, Watmann paper and nitrocellulose in transfer buffer (80% running buffer + 20% methanol)
2. Transfer protein electrophoretically using the transfer apparatus.

Steps for Developing Western Blot

1. Take the nitrocellulose paper and shake in the blocking agent for 1–2 hours.
2. Keep in TBS for 10 min (Tris-HCl pH 7.3 containing 0.15 m sodium chloride).
3. Transfer this paper to 1% primary antibody solution made in blocking agent. Keep at room temperature for 1 hour with constant shaking.
4. Rinse the paper in distilled water and keep in another container with TBS for 10 min. at room temperature.
5. Repeat step 3 with fresh TBS.
6. Add secondary antibody conjugated with alkaline phosphatase (1 : 2000 dilution) in 1% milk in TBS.
7. Shake for 1 hour at room temperature.
8. Rinse in distilled water and wash in TBS 10 min. twice.
9. Add the substrate solution (NBT + BCIP) in dark.
10. Color develops within ten min. take a photograph immediately.

Chapter **10**

Agarose Gel Electrophoresis (AGE)

10.1 INTRODUCTION

Agarose is a linear polymer composed of alternating residues of D and L-galactose joined by α (1 – 3) and β (1 – 4) glycoside linkage. The L-galactose residue has an anhydro bridge between the three and six positions. Chains of agarose form helical fibers that aggregate into supercoiled structures with a radius of 20–30 nm. Gelation of agarose results in a three-dimensional mesh of channels whose diameters range from 50 nm to > 200 nm.

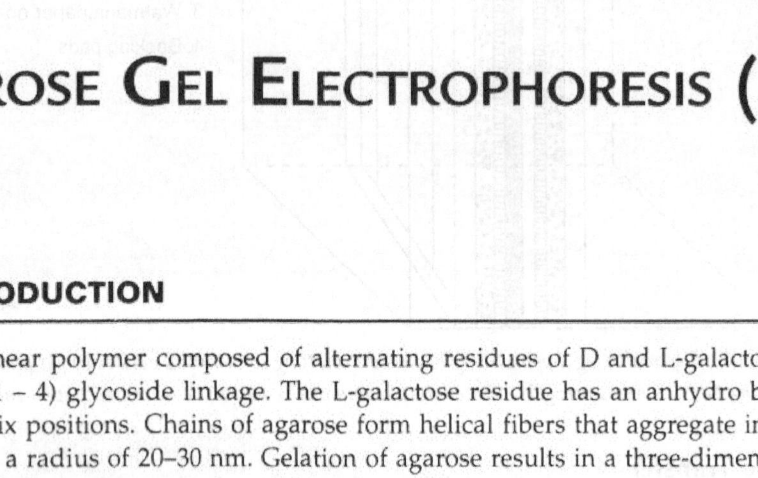

D-galactose 3,6-anhydro L-galactose

Commercially prepared agarose polymerase are believed to contain ~800 galactose residues per chain. However, agarose is not homogeneous, it varies from batch to batch and from manufacturer to manufacture. The presence of impurities and variability affact the gelling or melting temperature of agarose solution, the sieving of DNA, and the ability of the DNA recovered from the gel to serve as a substrate in enzymatic reactions. These potential problems can be minimized by using pure agarose purchased from authentic manufacturer.

The migration of DNA through agarose gel is affected by the molecular size of DNA, the concentration of agarose, the conformation of the DNA, the presence of ethidium bromide in gel and in buffer, applied voltage, the quality of agarose and the electrophoretic buffer. Molecules of double stranded DNA migrate through gel matrices at rates that are inversely proportional to the $\log_{10}$ of the number of base pairs. There is a linear relationship between the logarithm of the electrophoretic mobility of the DNA (μ) and the gel concentration (L) that is described by equation

$$\log \mu = \log \mu_0 - K_r L$$

where,

μ_0 = free electrophoretic mobility of DNA

K_r = the retardation coefficient (a constant related to the properties of the gel and the size and shape of the migrating molecules).

Some facts may be considered for best resolution of bands are as follows:

(*i*) The rate of migration of the linear DNA-dye complex through gels is consequently retarded by a factor of ~15%.

(*ii*) Migration is proportional to voltage applied. To obtain maximum resolution of DNA fragments > 2kb in size agarose gel should be run at no more than 5–8 V/cm.

(*iii*) The high ionic strength produces better resolution.

10.2 USE OF BUFFER

Electrophoretic Buffer

Several different buffers are available for electrophoresis of native, double stranded DNA such as Tris-acetate-EDTA (TAE), Tris-borate-EDTA (TBE) and Tris-phosphate-EDTA (TPE) at a concentration of ~50 mM (pH 7.5–7.8). All the buffers give satisfactory result and their use depend on only personal preference. The resolving power of TAE is slightly better than TBE and TPE.

Buffer	Working Solution	Stock Solution
TAE	1X	50X
	40 mM Trisacetate	242 g of Tris base 57.1 ml of glacial
	1 mM EDTA	acetic acid 100ml of 0.5 M EDTA
TPE	1X	10X
	90 mM Tris phosphate	108 g of Tris base 15.5 ml of
	2 mM EDTA	phosphoric acid 40 ml of 0.5M EDTA
TBE	0.5X	5X
	45 mM Tris-borate 1 mM	54 g of Tris base 27.5 g of boric acid
	EDTA	20 ml of 0.5 M EDTA

Gel Loading Buffers

Gel loading buffers are mixed with the samples before loading into the slots of the gel. These buffers serve three purposes

(*i*) they increase the density of the sample

(*ii*) ensuring that the DNA sinks evenly into the well

(*iii*) they add color to the sample

They are helpful in loading process and due to presence of dye in the electric field they move towards the anode at predictable rates. Bromophenol blue migrates through agarose gels ~2–2 fold faster than xylene cyanol FF. The migration of dye is independent of the concentration of agarose. The choice of loading dye depends on personal preferance.

Buffer Type	6X Buffer	Storage Temperature
I	0.25 bromophenol blue 0.25 xylene cyanol FF 40% (w/v) sucrose in H_2O	4°C
II	0.25 bromophenol blue 0.25% xylene cyanol FF 15% Ficoll in H_2O	Room Temperature
III	0.25% bromophenol blue 0.25% xylene cyanol FF 30% glycerol in H_2O	4°C
IV	0.25 bromophenol blue 40% (w/v) sucrose in H_2O	4°C

The migration of DNA during electrophoresis is also effected by pore size of gels, directly associated with gel concentration. Here a table represents the percentage of agarose standard for length of DNA fragments.

Agarose %	Standard	High Gel	Low Gelling/Melting Temp.
0.8	500 bp-15 kb	800 bp-10 kb	800 bp-10 kb
1.0	250 bp-12 kb	400 bp-8 kb	400 bp-8 kb
1.2	150 bp-6 kb	300 bp-7 kb	300 bp-7 kb
1.5	80 bp-4 kb	200 bp-4 kb	200 bp-4 kb

The important points for laboratory work:

(*i*) Position the comb 0.5 –1.0 mm above the base of tray so that a complete well is formed when the agarose is added to the mold.

(*ii*) The gel should be between 3 mm and 5 mm thick, check that no air bubble are under or between the teeth of the comb. Air bubble present in the molten gel can be removed easily by poking them with the corner of Kimwipe.

(*iii*) The maximum amount of DNA that can be applied to a well depends on the number of fragments in the sample and their sizes. The minimum amount of DNA that can be detected by photography of ethidium bromide stained gels is ~2ng in a 0.5 cm wide band. More sensitive dye such as SYBR Gold C can detect as little as 20 pg of DNA in a band. If there is more than 500 ng of DNA in a band of 0.5 cm, the slot will be overloaded, resulting in trailing, smiling and smearing problems that become more severe as the size of the DNA increases. When sample populations of DNA molecules are to be analysed, 100–500 ng of DNA should be loaded per 0.5 cm well. When the sample consists of a very large number of DNA fragments of different sizes, it is possible to load 20–30 µg of DNA per slot without significant loss of resolution. The dimension of the well is determined the maximum volume of solution that can be loaded. A typical well is of 0.5 × 0.5 × 0.15 cu cm and it contains about 40 µl. Do not over fill.

(*iv*) Close the lid of the gel tank and attach the electrical leads so that the DNA will migrate toward the positive anode (Red). Apply a voltage of 1–5 v/cm (measured as the distance between the positive and negative electrodes). The bubbles are generated in electrophoretic tank and within a few minutes, the bromophenol blue should migrate from the wells into the body of the gel. Run the gel until the bromophenol blue and xylene cyanol FF have migrated an appropriate distance through gel.

(*v*) During electrophoresis the ethidium bromide migrates towards the cathode (in the direction opposite to that of the DNA). Electrophoresis for prolonged periods of time can result in the loss of significant amounts of ethidium bromide from the gel, making detection of small fragments difficult.

The use of methylene blue is preferred than ethidium bromide due to protective measure, although it is less sensitive for DNA. In the condition of staining with methylene blue concentration of stain is increased 4–5 times for λ DNA and 2 times for plasmid DNA. A fluorescent light box for viewing slides and negatives provides ideal illumination for methylene blue gel. An overhead projector may also be used. Cover the surface of the light box or projector with plastic-wrap to protact the apparatus from liquid spills. A polaroid gun camera equipped with close-up diopter lens is used to photograph gel on either UV or white light transilluminator. Nowaday digital photography is becoming increasingly popular for gel documentation. Digital camera, especially those with close focus capability, can provide high quality image.

(*vi*) When the DNA samples or dyes have migrated to a sufficient distance through the gel turn off the electric current and remove the leads and lid from the gel tank.

(*vii*) Now observe the gel in UV light and photographed.

10.3 METHOD FOR LABORATORY

Agarose is a derivative of the agar-agar. This material is produced from the seaweeds. Agarose gel electrophoresis is an indispensable technique for DNA analysis.

1. Seal the ends of the gel casting tray with tape and insert a well forming comb. Place gel casting tray on the lab bench, so that agarose poured in next step can set undisturbed.
2. Prepare 0.8% (0.16 g in 20 ml water) agarose solution by heating. Allow to cool to 55–60°C. Add TEB to final concentration of 1X.
3. Carefully pour enough agarose solution into the casting tray to fill to a depth of about 5 mm. Gel should cover only about 1/3 rd the hight of comb teeth. Use a pipette tip to move large bubbles or solid deberies to the sides or end of the tray when the gel is still liquid.
4. Gel will become cloudy as solidified (in 10–15 minutes). Do not move or jar the casting tray when the agarose is solidifying. Touch corner of agarose away from the comb to test whether the gel has solidified.
5. When agarose has set unsealed the ends of casting tray. Place the tray on the plateform of gel box, so that comb is at negative black electrode (cathod).
6. Fills electrophoresis tank with 1X TEB. Remove the comb carefully. Make sure that the running buffer covers the entire surface of gel.

7. Load the wells with DNA samples mixed with 2 µl loading dye.
8. Connect the terminals and switch on the current. When dye reaches the edge of the gel switch off the electric supply.
9. Stain in 0.5 µg/µl (final/conc.) ethidium bromide for 10 min. Wash in water for 2 min.
10. Observe the gel in UV light and photograph.

Material and Instrumentation

- Electrophoretic Chamber and Comb
- Agarose
- Power Supplier
- Bromophenol blue/Methylene blue
- Running buffer
- Tank buffer

Loading Dye (2X)

• Bromophenol blue	0.05%
• TEB 10X pH 8.3	1 ml
• Glyceral	100 ul

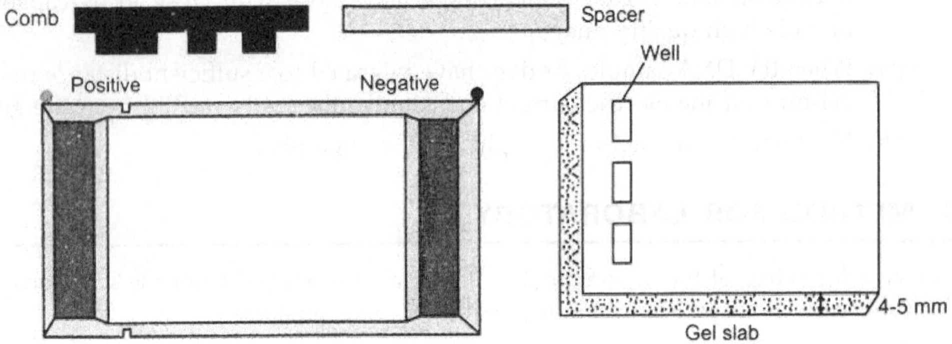

FIGURE 10.1 *Electrophoretic Instrument*

Running Buffer : Prepare 1X TEB

• TEB	10X (200 ml)
• Tris.HCl	21.6 gms adjust pH to 8.3
• EDTA	0.372 gms
• Boric acid	11 gms

Agarose ⟶ Solution ⟶ Melt ⟶ Solidity in tracking tray
↓

Put in electrophoretic tank after solidification pull comb gently, responsible for formation of wells for loading of sample.

Both sampling buffer and tank buffer ⟶ TBE Tris-barate-EDTA

Steps for Laboratory

A. Cast 0.8% Agarose gel

Seal the ends of gel casting tray with tap and insert a well forming comb. Place gel casting tray on the lab bench, so that agarose poured in next step can set undisturbed.

↓

Carefully poure enough agarose solution into the casting tray to fill to a depth of about 5 mm. Gel should cover only about 1/3 rd the hight of comb teeth. Use a pipette tip to move large bubbles or solid debris to the sides or end of the tray while the gel is still liquid.

↓

Gel will become cloudy as it solidifies (10–15 minutes). Do not move or jar the casting tray when the agarose is solidifying. Touch corner of agarose away from the comb to test whether the gel has solidified.

↓

When agarose has set, unseal ends of casting tray. Place the tray on the platform of the gel box. So that comb is at negative black electrode (cathod).

↓

Fill box with running buffer, to level that just covers the entire surface of the gel.

↓

Gently remove comb, taking care not to rip the wells

↓

Make sure that the sample wells left by the comb are completely submerged.

B. Working with Electrophoretic apparatus

(*i*) Running buffer is poured in platform.

(*ii*) Thickness of gel is determined by the amount of DNA/RNA in sample.

(*iii*) Solidification time is determined by the concentration by the concentration and thickness of gel. 1 gm/100 takes 15 min. 1.5 to 2% taken 45 min. of times.

(*iv*) Sample should be dissolved in minimum, mixed with sample buffer.

(*v*) Sample buffer

20% glycerol

2X TEB/TEA (Both samples and tank buffers are Bromophenol blue always same)
H_2O

(More volume of sample requires higher concentration of sample buffer)

(Some persons use phycol or sucrose instead of glycerol.)

(*vi*) The loading must be faster otherwise glycerol is dispersed in tank buffer.

(*vii*) Running buffer

TEB 10X : use 3 or 4 times

TEA 50X : use two times.

Pour is running tray for electrophoresis.

(*viii*) Connect tray with electrode to supply current. It is better to use low voltage. The generation of bubbles represents the initiation of process.

(*ix*) Disconnect power supply after proper observation.

(*x*) Observer gel in transilluminator and photographed.

(*a*) Staining

DNA–Ethidium bromide (2μg/ml) (orange in color but gel must not be colored) (always prefer to add in gel)

(*b*) DNA band is pencil like and RNA band is diffused.

Chapter **11**

Polyacrylamide Gel Electrophoresis (PAGE)

11.1 INTRODUCTION

Polyacrylamide gel electrophoresis is most effective for separation of small fragments of DNA (5–500 bp). Their resolving power is extremely high, and fragments of DNA that differ in size by as little as 1bp in length or by as little as 0.1% of their mass can be separated from one another. Although they can run very rapidly and can accommodate comparatively large quantities of DNA, polyacrylamide gel has the disadvantage of being more difficult to prepare and handle than agarose gels. Polyacrylamide gel is run in a vertical configuration in a constant electric field.

1. Prepare a standard denaturing sequencing gel; 4.75% acrylamide, 0.25% bisacrylamide 1X TEB, 7 M urea.
2. Pre run the electrophoresis gel for 30 min. at 60 W for a 30 × 40 cm gel.
3. Incubate the DNA and RNA to be electrophorized at 95°C for 5 min. and 65°C for 5 min. respectively.
4. Load the samples on the gel and run the electrophoresis at 40 to 50 W for a 30 × 40 cm gel until the bromophenol blue dye reaches the end of the gel and stain.

Solutions

1. 30% Acrylamide solution
   28.5 g Acrylamide
   1.5 g Bis-Acrylamide
   H_2O to 100 ml
2. 10% Ammonium persulfate
   Dissolve 1 g ammonium persulfate in 10 ml H_2O.
   Preferably make fresh each time, but can be kept for a week
3. TEMED (N,N,N',N'–Tetramethylendiamine)
   Store stock solution in cold.

11.2 METHOD FOR LABORATORY

Polyacrylamide Gel Electrophoresis (PAGE)

Take two pieces of glass plates and clean with alcohol or distilled water. Dry with tissue paper.

↓

Take small amount of grease and apply on both the side of spacer and placed on one plate. The greasing must be in one direction and uniform.

↓

Place spacer on three sides except notched boarder and kept another glass thereafter on it. To seal junctions by grease further. The thickness of spacer determines the thickness of gel.

↓

Now puts clips along spacer on three sides tightly.

↓

The level should be uniform. Now check the leakage by use of water on rim.

↓

Put separating gel through rim of small glass and let it for solidification except area for comb.

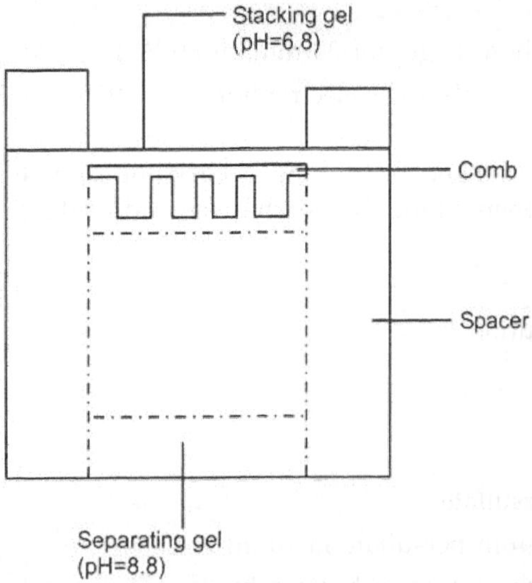

FIGURE 11.1

Fixed gel in running part with clips

↓

Fill running buffer in both upper and lower chamber with assurance of continuous current flow (voltage 70–80 mv, current 30 amp)

↓

Now set comb after putting staking gel.

↓

After sometime (solidification) pull comb most gently to create filling wells.

↓

Fixed gel in running part with clips.

↓

Fill wells with samples under study.

↓

Supply a continuous current flow for band migration.

↓

Observe gel under UV transilluminator and photographed by close image digital camera or other photographic camera has UV filter.

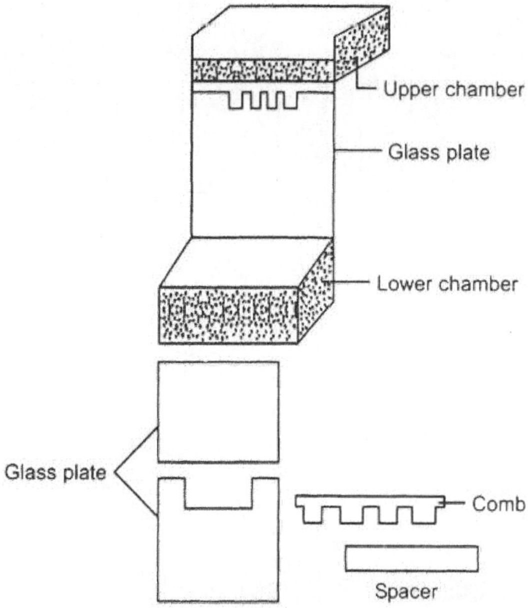

FIGURE 11.2

Bottom or Separating Gel

Acrylamide	=	2 ml
Separating gel buffer	=	2.5 ml
Water	=	5.5 ml
Mix well and add SDS	=	50 µl

↓

Then both TEMED (30 µl) and ammonium per sulphate add simultaneously. Pour immediately since, polymerization is initiated.

↓

The solution is 12.5% gel with pH 8.8.

Stacking Gel

• Acrylamide	=	1 ml
• Staking gel buffer	=	1.25 ml
• Water	=	2.5 ml
• SDS	=	20 µl
• TEMED	=	10 µl (20 µl)
• Ammonium phosphate	=	20 µl (30 µl)

The solution is 6% gel with pH = 6.8.

Chapter 12

RECOVERY OF DNA

12.1 INTRODUCTION

Plasmid or genomic DNA fragment can be recovered by electroelution or by phenol extraction for cloning specific fragments of DNA after restriction digestion with and for preparing probes for hybridization.

Although many methods have been developed over the years to recover DNA from gels, none have proved entirely satisfactory. The major problem is the contamination of agarose with sulphated polysaccharide, which is extracted from the gel together with DNA. These substances are potent inhibitors for many of the enzymes (*e.g.*, restriction endonucleases, ligases, polymerases etc.) that are commonly used in the subsequent cloning. Therefore a good quality agarose should be used for the laboratory purpose. The efficiency of the DNA extracted from agarose gel is the function of its molecular weight.

As the molecular weight of DNA increase the yield change steadily. So that the fragments greater than 20 kb in size is rarely recovered.

12.2 RECOVERY OF DNA FROM AGAROSE GEL

The methods used for the recovery of DNA from agarose gel are:

- Electroelution
- Freezing thawing methods
- Homogenizing method
- Low melting agarose method
- Agarose enzyme treatment

I. Electroelution

1. Cut a ethidium bromide stained band from the agarose gel under UV.
2. Put the cubes of agarose gel into a dialysis bag filled with sterile electrophoresis buffer (TEB 0.5X).

3. Squeeze out most of the electrophoresis buffer. Leave no air bubbles.
4. Place the gel on a horizontal electrophoresis tank containing a shallow layer of TEB (1X cold).
5. Electrophores at 100 volts for 2–3 hours.
6. Reverse polarity for 1 min.
7. Remove all liquid from the bag and rinse it.
8. Examine the gel slice under UV. All DNA should be eluted.
9. Combine the fluids from the bag. Reduce the volume and remove ethidium bromide by butanol.
10. Extract with phenol/chloroform and chloroform.
11. Add 3 M sodium acetate (0.1 volume). Precipitate DNA with ethanol.
12. Centrifuge and dissolve DNA in TE.

Laboratory Steps

Cut a piece of dialysis membrane and wash with sterilized distilled water

↓

Taged one end with cord and filled runnig buffer (TEB 2X)

↓

Add a piece of agarose gel containing DNA.

↓

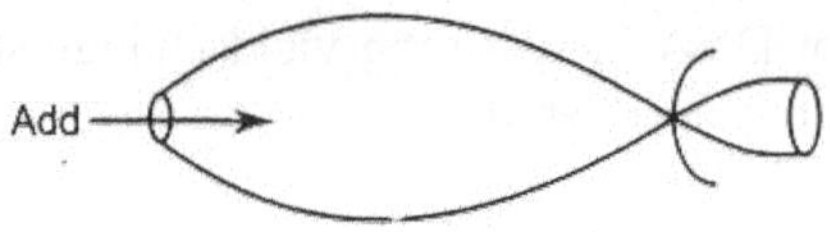

FIGURE 12.1

Place the dialysis perpendicularly in running tray of electrophoresis and fixed its position by the use of clip tag.

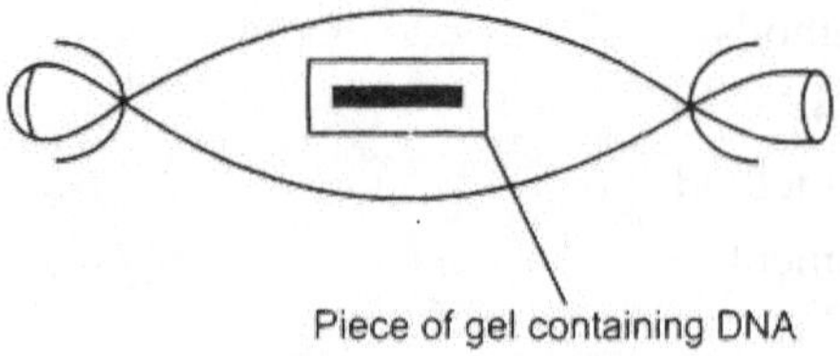

FIGURE 12.2

Electrophorse at 100 volts for 2–3 hours

↓

Reverse polarity of current for very short time.

↓

Cut one tied end and suck buffer with the help of micropipete

↓

Keep all buffers in two/three eppendorfs

↓

Add equal amount of butanol and mix gently.

↓

Centrifuged 2000 rpm for 3 min.

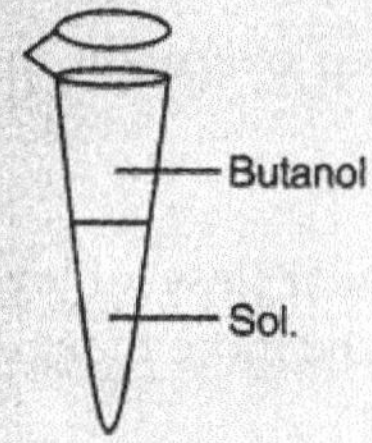

Keep in other eppendorf

↓

Repeats four times butanol treatment.

↓

Add 2 vol of alcohol

↓

Keep over night in freezer

II. Freezing Thawing Method

Keep a piece of gel containing DNA in eppendorf.

↓

Add saturated phenol to dissolve and break the gel.

↓

Keep in freezer over night

↓

Centrifuged 3000 rpm for 5 min.

↓

Supernatant (DNA solubilised material) is collected in other eppendorf.

↓

Add equal volume of chloroform and mix for two minutes.

↓

Centrifuged 3000 rpm for 2 min.

↓

Separate supernatant in other eppendorf and mix ethanol (1 ml)

↓

Keep overnight in freezer

III. Homogenizing Method

1. Electrophorese the DNA in 1% or (less) of low melting temperature agarose/acrylamide. (Warning) : power should not be more than 3A or 7.5 v/cm to prevent melting of agarose during electrophoresis.
2. Stain the gel. Cut out the needed band and transfer it to an eppendorf tube.
3. Keep the tube at 70°C for 3 min.
4. Add 0.1 volume of 5 M NaCl and vortex.
5. Keep at 70°C for 3 min.
6. Extract the solution 2–4 times with equal volume of buffered phenol.
7. Extract twice with chloroform : Isoamylalcohol.
8. Extract the aqueous phase with butanol several times to remove ethidium bromide and to reduce the volume.
9. Add 2.5 vol ethanol. Keep at –20°C for 2 hours.
10. Spin 10000 rpm for 10 min.
11. Dissolve DNA in 20 μl sterile distilled water or TE.
12. Repeat step 9 to remove the salts, if needed.

Cut a piece of gel has DNA and keep it in an eppendorf.

↓

Homogenized the piece of gel

↓

Add phenol and mix it

↓

Centrifuged 3000 rpm for 5 min.

↓

Collect supernatant in other eppendorf.

↓

Followed by chloroform alcohol treatment (Chapter-4).

IV. Low Melting Agarose Method

After slow heating upto 45°C, it is followed by phenolo-chloroform treatment.

V. Agarase Enzyme Treatment

The piece of gel containing DNA is treated by enzyme agarase. But it is cost intensive.

(*i*) Before use ethidium bromide is removed by treatment with butanol. But it is most sensitive work since butanol is present on upper layer and material is present in lower layer.

(*ii*) In some labs 20 X SSC saturated isopropanol is used. But it is not good method.

Chapter 13

SILVER STAINING

13.1 INTRODUCTION

The detection of DNA fragments in polyacrylamide gels by ethidium bromide staining is less sensitive than in agarose gels. Also ethidium bromide is highly carcinogenic and a trans illuminator is required for viewing. In order to avoid exposure to UV radiation and carcinogenic ethidium bromide, silver staining is advised.

Silver staining is a highly sensitive method for detection of nucleic acids in polyacrylamide slab gels. It is not recommended for agarose gels. Both double-stranded DNA and single-stranded DNA and RNA can be detected by silver staining. The principle of silver staining has been based on the work of *C.A Von Bollstadt* (1906–1980) recorded that silver nitrate would stain human skin. Later on silver staining has been refined into a highly sensitive technique for post electrophoretic detection of DNA bands in agarose and polyacrylamide gels. At its best, silver staining can detect bands containing < 1.0 ng DNA. The three general types of silver straining are as follows:

1. Photo development uses photonic energy to reduce silver ions to the metallic element. Unfortunately, the simplicity and speed of the method cannot compensate for its lack of sensitivity, which is better than be achieved by conventional staining with ethidium bromide.
2. Diamine staining methods use ammonium hydroxide to generate silver diamine complexes, which bind to the nucleic acid. Silver ions are then liberated from the complexes by decreasing the concentration of ammonium ions with litnic acid. The liberated silver ions finally reduced to metallic silver by formaldehyde. The basic method is both rapid and sensitive on a regular basis. In addition, ammonical silver salts are potentially explosive and must be handed with great care.
3. Non-diamine staining involves fixation of the DNA, sensitization of the DNA with glutaraldehyde, impregnation of the gel or membrane with silver nitrate at weakly acidic pH, and reduction of bond silver ions to metallic silver by alkaline formaldehyde. Many different variants of this technique have been published, but most of them suffer to a greater or lesser extent from the same problem ; bands of DNA stain gray or dark yellow against a variable background of brownish surface staining. This problem can be

minimized by carefully monitoring of the gel during development so as to obtain the greatest discrimination between specific staining of the gel. Differential reduction of silver ions can be improved by adding sodium thiosulphate to the alkaline formaldehyde solution. Thiosulphate removes silver ions from the gel surface by forming soluble complexes with silver salts. The non-diamine staining detect bands of DNA containing 2–5 ng DNA.

13.2 METHODS FOR LABORATORY

Material and Instrumentation (DNA)

- Acetic Acid (3% v/v)
- Developer
- Ethanol (10% v/v)
- Ethanol/Glycerol (10% v/v / 7% v/v)
- Formaldehyde (37 v/v)
- Nitric Acid (0.7% v/v)
- Silver nitrate fresh (0.2% w/v)

Procedure

After electrophoresis is completed, placed the gel in a plastic tray for silver staining

↓

Remove the buffer by washing the gel in distilled water

↓

Keeps gel in 10% Ethanol for 10 min. necessary for fixation, Remove ethanol from tray

↓

Keep gel again in 10% ethanol for 5 hour and remove ethanol from tray.

↓

Take 7% nitric acid to cover the gel, shake the gel for 6–8 minutes. Remove nitric acid by washing of gel with distilled water twice.

↓

Take 0.2% silver nitrate to cover the gel. Shake the gel gently for 30 minutes. Wash gel with distilled water thrice.

↓

Take 100 ml developer and add 125 µl of formaldehyde solution. Shake the tray gently in indirect light. When the solution turns yellow or when a dark precipitate becomes noticeable, replace solution by another 100 ml of developer plus 125 μl formaldehyde solution. Continue to shake the gel in indirect light.

↓

Observe the appearance of bands and background. Remove developer solution.

↓

Take 250 ml of 3% acetic acid in tray. Shake the gel gently for five minute. Remove acetic acid.

↓

Take 10% ethanol and repeats the further change by 10% ethanol (2 min.)

↓

Keep the gel in 10% ethanol/7% glycerol for long-term storage.

Material and Instrumentation (RNA)

- Methanol (50%)
- Acetic Acid (7%)
- Gluteraldehyde (5%)
- Sodium thiosulphate (5%)
- Silver nitrate (0.1%)
- Formaldehyde
- Sodium carbonate (3%)
- Citric Acid (2M)

Can be used for RNA/DNA/Protein.

For Ribonucleic Acid

Place gel in tray after electrophoresis for silver staining

↓

10 min. in fixatives 50% methanol, 7% Acetic acid

↓

Wash in distilled water

↓

Keep in 5% gluteraldehyde for 10 min. if sample is DNA or RNA. The 5% Sodium thiosulphate is used for protein.

↓

Wash in distilled water

↓

Keep in 0.1% Silver nitrate for 5 min.

↓

Wash in distilled water

↓

Rinse and put in 50 µl of 37% formaldehyde (developing immediate) 100 µl of 3% Sodium carbonate for color (develops)

↓

Stop by the use of 2.3 M citric acid for 5 min.

↓

Store in water

Chapter **14**

BASIC CYTOGENETIC TECHNIQUES: STUDY OF CHROMOSOME

14.1 INTRODUCTION

The study of chromosome is first hand necessity to initiate work on cytogenetic. The best stage for karyotypic study and preparation of ideogram is metaphase. The laboratory method for both animals and plants are described in present book.

14.2 METHOD FOR BLOOD

Material and Instrumentation

- Fetal Bovine Serum (FBS)
- RPMI 1640 Medium
- L–glutamine
- Gentamicin (50 mg/ml)
- Phytohemogglutinin-M (PHA)
- Colcemid (10 μg/ml)
- Hank's Balanced Salt Solution (HBSS)
  - — Amethopterin
  - — Thymedine
  - — Potassium Chloride
  - — Ethidium Bromide
  - — Sodium Chloride
- Sodium heparin Vacutainer
- One end frosted slide (should be hydrophilic)
- Tuberculin syringe

- Conical centrifuge tube (15 ml)
- 21–gauge niddle
- trypsin
- Wright stain
- Whatmann 10–1 filter paper
- Mounting medium
- pHydrin capsule
- A table—top centrifuge
- Phase—contrast microscope

Procedure

Collect peripheral blood into sodium heparin vacutainer. The blood can be kept for 5 days before culture. The blood is collected by 21 gauge niddle and cultured in 5.0 ml RMPI 1640 medium.

15 ml centrifuge tube (sterile)

↓

Add 0.05 ml of PHA solution

↓

Tightly cap the tube and incubate culture at a shallow angle at 37°C for 3 days.

↓

21 hours prior to harvest, block the culture by adding 50 μl of 10^{-5} M amethopterin. Further incubate at 37°C.

↓

4½ hours prior to harvest, release the culture by adding 50 μl of 10^{-5} M thymidine. Further incubate at 37°C.

↓

30 minutes prior to harvest add colcemid to a final concentration 0.05 μg/ml. Further incubate at 37°C.

↓

One hour before harvest add 50 μl of 1.25×10^{-3} M ethidium bromide to culture.

↓

10 min. prior to harvest add colcemid and incubate at 37°C

↓

Start harvest by centrifuging at a relative centrifugal force (180 g for 8 min. and aspirate supernatant.)

↓

Add 6 ml of 0.075 M KCl hypo-tonic solution at room temperature to the tube and gently re-suspend. Incubate for 15 min. at room temperature.

↓

Add 10–12 drops of 3:1 methanol : acetic acid fixative using pasture pipette. Mix gently by inversion of tube.

↓

Centrifuge the tube. The pellet is observed appears brown and clumpy.

↓

Remove supernatant except 0.5 ml. Re-suspend the pellet and add 5 ml of fixative slowly. Mix well by inverting the tube and centrifuged. Repeat the process.

↓

Remove the supernatant and add appropriate volume of fixative for slide making (light milky suspension), take cell suspension in pasteur pipette. Hold pipette horizontally about 1–2 inch above the side and place 3 drops of suspension on to the slide starting from the free end and moving towards the frosted end. The drops should strike the slide from the top and should burst on the water film as they strike the slide and spread out evenly.

↓

Flood the slide by applying fixative across the top of the slide from the free end towards the frosted end.

↓

Blot the long edge of the slide in a straight angle on the paper towel, wipe off the back and air dry the slide. The drying process can be manipulated depend on humidity and temperature of environment.

↓

The contrast of the chromosomes should be observed by phase contrast microscope.

↓

Slides are either baked at 60°C for 2 hrs or aged overnight at room temperature.

↓

Dip the slide in trypsin solution for 60 sec.

↓

Rinse briefly in two coplin jars containing 0.9% NaCl.

↓

Flood the slide for 1–2 min with a mixture of 1 ml of 0.2% Wright stain and 4 ml of pHydrain working solution

↓

Rinse with H_2O and blow dry .

↓

Cover the slide with mounting medium and cover glass.

Processing of Slide

The processing of slide before its use for the preparation of chromosome spread is also an important step. The blood cells burst on slide due to the hydraulic pressure created by water film. At first slides are kept in a coplin jar containing 100% methanol. The slides are removed and their one side is polished by kimwipe. The slides are again rinse in 100% menthol and transferred in a beaker containing deionized water. The slides are removed when a uniform thin film of water covers the surface. Now slides are kept in hand their long edge is wiped out by a paper towel to remove excess of water.

Solution Preparation

1. **Culture Medium RPMI 1640**
   RPMI 1640 medium = 100 ml
   Fetal bovine serum = 11 ml
   200 mM L-glutamine = 1.2 ml
   50 mg/ml gentamycine = 0.1 ml
2. **Amethopterin Stock Solution (10^{-5} M)**
   Amethopterin = 0.5 mg
   Hank's BSS = 98 ml
   Aliquots can be stored at 4°C for one month.
3. **Thymidine Stock Solution (10^{-3}M)**
   Thymidine = 25 mg
   Hank's BSS = 100 ml
   Aliquots can be stored at 4°C for one month.
4. **Hypotonic Solution (0.075 M KCl)**
   Potassium chloride = 0.56 gm
   Sterilized D. water = add to make 100 ml
5. **Fixative (Fresh)**
   Methanol = 30 ml
   Glacial acetic acid = 10 ml

6. **Ethidium Bromide (1.25 × 10^{-3} M)**

   Ethidium bromide = 4.9 mg

   Hank's BSS = 10 ml

   Protect the solution from light with foil and store at 4°C.

7. **Trypsin**

   Rehydrate vial with 10 ml H_2O and store 1 ml aliquots at – 20°C. Mix 1 ml of the trypsin solution and 49 ml of 0.9% NaCl before use.

8. **HBSS**

   Sodium chloride = 8 gm/l

   Potassium chloride = 400 mg/l

   KH_2PO_4 = 60 mg/l

   Na_2HPO_4 = 47.86 mg/l

   Glucose = 1000 mg/l

   $NaHCO_3$ = 350 mg/l

   Autoclave to sterilize. In one liter of distilled water.

9. **Wright's Stain (0.2%)**

   Wright stain = 0.2 gm

   Methanol absolute = 100 ml

   Stir and mix well, filter through whatman paper no-1 and store in brown bottle, discared after one month.

10. **pHydrion Solution**

    Dissolve the contents to one pHydrion 6.2 capsule in 100 ml of distilled water to make a stock.The working solution is 5% of the stock solution.

11. **Sodium Chloride Solution**

    Dissolve chloride = 0.9 gm

    Distilled water = 100 ml

14.3 METHOD FOR PLANTS

Material and Instrumentation

The study of chromosome is occurred by smearing technique in plants. The method consists spreading of the cells in a single layer.

- Microsporocytes/Anthers/root tip/apical merismatic tissue
- Sulphuric acid-potassium bicarbonate mixture, Navastin fluid
- Acetocaramine/Giemsa stain
- Sprit lamp
- Slide
- High resolution microscope

Procedure

A clean slide is immersed for 2 hour in sulphuric acid-potassium bicarbonate mixture then wash in water and blot to dry.

↓

Place fresh tissue on slide(dissected if necessary.)

↓

Smear the tissue by the use of other slide carefully.

↓

Slide is inverted over a petri disc containing Navashin fluid for 10–15 min.

↓

Slide is inverted with smear at upper side.

↓

A few drop of acetocaramine/acetoorecin/giemsa stain are placed on smear for 5 min. Stain is replaced with a fresh drop of stain.

↓

After crushing the tissue mass debris are removed.

↓

Slide is gently heated over a flame and cover glass is placed on the material. A uniform pressure is applied on the material by placing blotting paper on the cover glass and the pressing it.

↓

Slide is immediately sealed with melted wax.

Solution Preparation

1. Navashin Fluid

Solution A: Chromic anhydride – 1.5 ml
Glacial Acetic Acid – 10 ml
Distilled water – 90 ml

Solution B: 40% Formalin – 40 ml
Distilled water – 60 ml

Both solutions are mixed in 1:1 ratio just before use.

2. Acetocaramine

Glacial acetic Acid - 45 ml

Distilled water - 55 ml

Heat the solution to the boiling point and add 1 gm of carmine powder.

Cool and filter the solution.

3. Acetoorecin

Commercially available in market.

4. Giemsa stain

The Giemsa stain should be freshly prepared in phosphate buffered saline in H_2O and filtered through Watmann no. 1 filter paper before use.

Chapter **15**

BACTERIAL CULTURE

15.1 INTRODUCTION

In the experiments associated with recombinant DNA preparation, the knowledge of bacterial culture is significantly important. The bacterium commonly used for culture is *Escherechia coli*. It is a commensal organism of Homo sapiens, and a normal composition of bacterial fauna of the human gut. It is non-pathogenic and rarely associated with any type of disease. The culture of *E.coli* is easy because it has simple nutritional requirement and grow slowly on "minimal" medium containing

- an energy source such as glucose
- salts such as NaCl and $MgCl_2$
- the vitamin biotin
- the vitamin thiamine (B_1)

E.coli synthesizes all necessary vitamins and amino acids from these precursors and grows rapidly in a complete medium, such as LB. Yeast extract and hydrolyzed milk protein (casein) provide a ready supply of vitamins and amino acids.

Pattern of Growth

The bacterium represents three phases of growth in closed condition of nutrient supply or cultural medium. The different phases of growth are referred as lag phase or quiescent phase, log phase or exponential phase and stationary phase. However some workers also included death phase or decline phase. In lag phase bacterial cell adjusts in new environment and represents growth mainly in volume and not in number. It lasts in 30–60 minutes following inoculation. In second phase, it divides rapidly and represents growth in number. Each generation and division is responsible for doubling of number, known as doubling time. The doubling time of bacterium *E.coli* is 25–30 minutes. As the nutrients of cultural media is depleted, the growth rate is decreased and finally cells stop division (carrying capacity). During this phase the number of bacteria remain constant, known as stationary phase. The phase is succeeded by decline or death phase due to accumulation of more and more excretory products and anti-metabolites.

Optimum growth in liquid cultural medium is achieved with continuous agitation, which aerates the cells, facilitates the exchange of nutrients and flushes away waste products of metabolism. It can be safely assumed that a culture in complete medium has reached stationary phase following overnight incubation with continuous shaking.

A = Lag phase

B = Log phase

C = Stationary phase

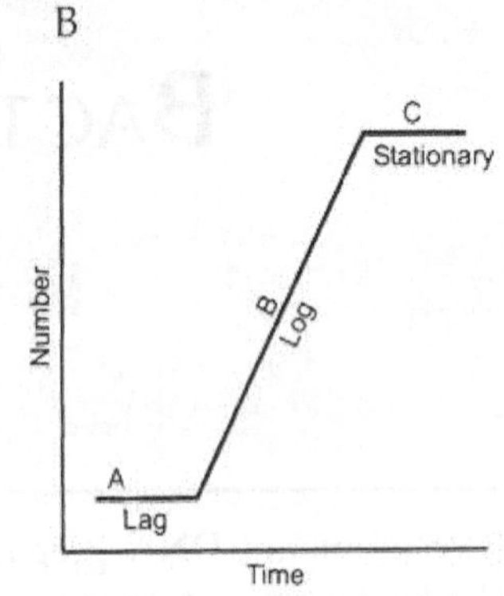

FIGURE 15.1. *Growth pattern of bacteria in closed cultural medium*

15.2 PLATE STREAKING CULTURE

The method is used for preparation of visible colony from a single isolated cell. Streaking cells to obtain individual colonies is usually the first step in genetic manipulation of microorganisms. Using cells derived from a single colony minimizes the chance of using a cell mass contaminated with foreign microorganisms. To demonstrate antibiotic resistant, the growth of wild type *E.coli* and that of an *E.coli* containing an ampicillin resistance gene are compared, using LB medium containing ampicillin. The resistant strain contains the plasmid pAMP, which produces an enzyme (β-lactamase) that destroys the ampicillin in the medium, thus allowing these cells to grow.

Material and Instrumentation

- LB agar
- Ampicillin plates
- Refrigerator
- *E.coli*
- Inoculating loop
- Incubator
- Boiling Water Bath
- Sprit lamp

Before the initiation of plate streaking, we prepare LB agar plate and LB/Amp plate. The pre-sterilized agar is melted in boiling water bath or micro-oven, cooled to approximately 60°C, and poured into sterilized cultural plates. It is precautionary measure to take half filled container to prevent boiling over and loosen the cap to prevent bottle from exploding. The antibiotic ampicillin is added when agar cools upto 50°C to prevent neutralization of antibiotic due to heat, since antibiotics are thermolabile chemicals. Ampicillin is very stable in agar plates, thresholds for selection are relatively broad and contaminants are infrequent.

Procedure

Mark plates by permanent marker as LB plates and LB/Amp plates. One plate from both is marked as –pAMP for cell without plasmid and + pAMP for cell with plasmid.

↓

Take the inoculating loop and sterilize it by heating in sprit lamp and cool it. To avoid contamination do not place inoculating loop on lab bench.

↙ ↘

FROM CULTURAL PLATE

At first stab inoculating loop into a clear area of the agar several times to cool . Use the loop tip to scrape up a visible cell mass from a colony. Do not gouge agar.

FROM STAB CULTURE

Remove cap of vial but avoid touching rim of cap. Quickly pass mouth of vial several times through burner flame. Scrape the loop several times in area of culture where bacterial growth is apparent. Remove loop, flame vial mouth and replace cap.

↓

Take plate one by one and streak the culture plates-

Streak–1 : Glide inoculating loop tip back and forth across the agar surface to make a streak across the top of the plate. Avoid gouging agar. Replace lid of plate between streak.

Streak–2 : Re-flame inoculating loop and cool by stabbing it into the agar away from the first (primary streak). Draw loop tip through the end of the primary streak and, without lifting loop, make a zigzag streak across ¼ of the agar surface.

Streak–3 : Re-flame inoculating loop and cool again. Draw loop tip through the end of the secondary streak and make another zigzag streak in the adjacent quarter of the plate without touching the previous streak.

Streak–4 : Sterilize inoculating loop by heating. Draw tip through the end of tertiary streak and make a final zigzag streak in remaining quarter of plate. Keep inoculating loop after sterilization by heating. Develop a habit for such type of activity.

↓

Place plates upside down in a 37°C incubator and incubate for 15–20 hours. Plates are inverted to prevent condensation that might collect on the lids from falling back on the agar and causing colonies to run together.

↓

Optimal growth of well-developed individual colonies is achieved in 15–20 hours. At this point colonies should range in diameter from 0.5 mm to 3 mm.

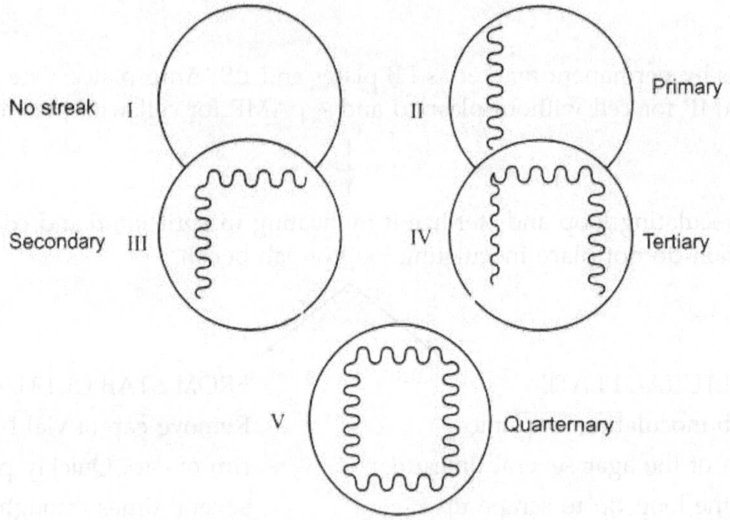

FIGURE 15.2 *Plate Streaking Culture*

Antibiotic Resistant Growth

If the plates cannot be observed on the day after streaking, store them at 4ºC to arrest *E.coli* growth and to slow the growth of any contaminating microbes. Wrap in parafilm or plastic wrap to retard drying. Observe plates and use the matrix below to record which plates have bacterial growth and which have no growth. On plates with growth, distinct individual colonies should be observed within one of the streak. On the LB/Amp plate, growth must be observed in the secondary streak to count as antibiotic resistant growth. In a heavy inoculam, non-resistant cells in the primary streak may be isolated from the antibiotic on a bed of other non-resistant cells.

	Transformed cell +pAMP	Wild type cell–pAMP
LB/Amp	Experiment	Negative control
LB	Positive control	Negative control

On the LB/Amp +pAMP plate, tiny satellite colonies may be observed radiating from the edges of large well established colonies. These satellite colonies are not ampicillin resistant, but grow in an antibiotic shadow, where ampicillin in the media has been broken down by the large resistant colony. Satellite colonies are generally a sign of antibiotic weakened by not cooling the medium enough before adding the antibiotic, long-term strong of more than 30 days, or over incubation.

15.3 OVERNIGHT SUSPENSION CULTURE

It is a method for growing small-scale suspension culture of *E.coli.* In this, bacterial cells are reached in stationary phase. It is commonly used for purification of plasmid DNA and for inoculating mid log culture. The bacterial cells use for overnight suspension culture is selected from streak culture of plate LB/Amp agar. Although it is always best to grow an overnight culture with shaking, it is not essential when growing cells for purifying plasmid or for inoculating a larger culture. For these purpose, suspension can be incubated, without shaking, in a rack within a 37°C incubator. However, the cultures will need to incubate for 1 day or more to obtain an adequate number of cells.

A 50 ml conical tube is preferable for growing overnight cultures. It provides greater surface area for aeration than does a 15 ml culture tube, although a 15 ml tube can also be used. In overnight culture for plasmid preparation, it is best, but not essential to maintain antibiotic selection of the transformed strain by growing in medium containing the appropriate antibiotic. It is prudent to inoculate an earlier overnight culture, in case first was not inoculated properly.

For plasmid purification we commonly use freshly streaked plate (within week) whereas for mid log culture, it is necessary to make the overnight culture exactly the size of the inoculam needed to start mid log culture; 1 ml overnight per 100 ml mid log culture.

Material and Instrumentation

- *E.coli* plate
- Conical tube (50 ml)
- Shaking water bath
- LB broth
- Pipette (10 ml)
- Bleach (10%)
- Sprit lamp
- Inoculating loop

Procedure

Take a sterile 50 ml tube, the large tube provides a greater surface area for good aeration of culture.

↓

Use a 10 ml pipette to sterilely transfer 5 ml of LB broth into tube.

↓

Locate well defined colony 14 mm in diameter on a freshly streaked plate.

↓

Sterilize inoculating loop in the burner flame until it glows red hot, then continue to pass lower half of its handle through the flame.

↓

Cool loop tip by stabbing it several times into agar near the edge of plate.

↓

Use loop to scrape up a visible cell mass from selected colony.

↓

Sterilely transfer colony into culture tube. Briefly re-flame mouth of culture tube and replace cap. Re-flame inoculation loop before placing it on lab bench.

↓

Loosely replace cap to allow air to flow into culture. Affix a loop of tape over the cap to prevent it from becoming dislodged during shaking.

↓

Incubate for 12–24 hours at 37°C, preferably with continuous agitation. Shaking is not essential for a culture to be used for plasmid purification. The culture can be incubated at 37°C, without shaking for 1 or more days.

Method for Calculation of Concentration

The method for calculation of concentration of cell in original culture is based on serial dilution. We take ten tests tube filled with 1 ml of LB medium. In first tube we take 10 µl overnight culture and serially diluted the culture up to tenth test tube. From tenth tube we take 100 µl and spread on agar plate. Incubate plate overnight at 37°C and count the number of colony developed. Suppose we observed 20 colonies.

Since 100 µl produces 20 colonies of bacterium *E.coli.*

1000 µl produces (20 × 1000) ÷ 100 = 200

The tenth tube is 10 ×10 fold dilution of original culture. So the number of bacteria in original culture is 200 (10 × 10) per milliliter.

15.4 MID LOG SUSPENSION CULTURE

The mid log suspension culture is prerequisite for preparation of competent cell. Commonly the ratio of mid log cells to competent cells is 10 to 1. For example, 10 ml of competent cells are derived from 100 ml of mid log culture. During processing of culture absolute degree of sterility is maintained because any contaminant will multiply as cells are repeatedly manipulated and/ or stored for future use.

A shaking incubator is necessary for growing *E.coli* for competent cells. Proper aeration and nutrient exchange are essential to achieve vigorous growth, only cells collected during the middle part of log (mid log) phase will provide competent cells with a high transformation frequency. An economical alternative to a shaking water bath or temperature controlled dry shaker is to place a small platform shaker inside a 37°C incubator. The adequate surface to volume ratio is necessary for the exchange of air and cells should be grown in an Erlenmeyer flask with a volume of LB broth not exceeding one third of the total volume of the flask.

Incubate a mid log culture 2–4 hours before experiment for transformation begins. *E.coli* strain reaches mid log phase after 2–2.5 hours of incubation. Cells can be used immediately or held on ice for up to 2 hours before beginning of transformation. Timing of the culture to reach mid log phase is likely to be affected by any change in physical and laboratory conditions. For example, a culture incubated with an overnight culture that was grown without shaking will take longer time to reach mid log phase. A culture begin by inoculating into LB broth pre-warmed to 37°C will reach mid log phase more quickly than one begin at room temperature. Different strains of *E.coli* display different growth properties.

Meterial and Instrumentation

- Overnight suspension culture
- LB broth (Sterile)
- Sprit lamp/Bunsen burner

- Erlenmeyer flask (500 ml) sterile
- Pipettes (10 ml) sterile
- Shaking water bath
- Spectrophotometer

For sterilization autoclave glass wares 15 minute at 121°C, cool and store at room temperature. Add 100 ml of LB broth to a 500 ml flask and cap with aluminum foil as a microbe barrier.

Procedure

Sterilely transfer 1 ml of overnight culture into 100 ml of LB broth at room temperature. Remove cap from overnight culture tube and flame tube. Do not place cap on lab bench. Remove foil cap from flask and flame mouth. Do not place cap on lab bench. Pour overnight culture into flask. Reflame mouth of flask and replace foil cap. At first flame pipette cylinder if only a portion of overnight culture is required. Remove cap from overnight culture tube and flame mouth. Do not place cap on lab bench. Withdraw 1 ml of overnight suspension. Remove foil cap from flask and flame mouth. Expel overnight sample into flask. Reflame mouth of flask and replace foil cap.

↓

Incubate 37°C with continuous shaking

↓

Approximately 1 hour after inoculation, sterilely withdraw a 1 ml sample of the culture and measure absorbance (optical density at 550 nm). Repeat procedure at approximately 20 minutes interval and draw a graph. The OD increases 0.3–0.5 in 90 to 100 minutes. Store mid log culture on ice until ready to begin the process of transformation.

APPENDIX–1

COMMON SOLUTIONS AND MEDIA

LURIA BERTANI MEDIUM (LB)

- Deionized H_2O = 950 ml/litre
- Tryptone = 10 gm/litre
- Yeast extract = 5 gm/litre
- NaCl = 10 gm/litre

Shake until the solutes have dissolved. Adjust the pH to 7.0 with 5 N NaOH (~0.2 ml). Adjust the volume of the solution to 1 liter with deionized H_2O. Sterilize by autoclaving for 20 minutes at 15 psi (1.05 kg/cm^2) or liquid cycle.

TERRIFIC BROTH (TB)

- Deionized H_2O = 900 ml/litre
- Tryptone = 12 gm/litre
- Yeast extract = 24 gm/litre
- Glycerol = 4 gm/litre

Shake well to dissolve solute and sterilize by autoclaving for 20 minutes at 15 psi (1.05 kg/cm^2) or liquid cycle. Heat solution up to 40°C and cool. Add 100 ml of a sterile solution of 0.17 M KH_2PO_4, 0.72 M K_2HPO_4.

0.17 M KH_2PO_4, 0.72 M K_2HPO_4

- KH_2PO_4 = 2.31 gm
- K_2HPO_4 = 12.54 gm
- Deionized H_2O = 90 ml

Shake well to dissolve and adjust the volume of the solution to 100 ml with desired H_2O and sterilize by autoclaving for 20 minutes at 15 psi (1.05 kg/cm^2) on liquid cycle.

SOC MEDIUM

Sol. A :	Deionized H_2O	= 950 ml/litre
	Tryptone	= 20 gm/litre
	Yeast extract	= 5 gm/litre
	NaCl	= 0.5 gm/litre

Shake well to dissolve solutes. Add 10 ml of a 250 mm solution of KCl (1.86 g KCl in 100 ml of deionized H_2O). Adjust the pH of the medium to 7.0 with 5 N NaOH (0.2 ml). Make volume of the solution to 1 liter with deionized H_2O. Sterilize by autoclaving for 20 minutes at 15 psi on liquid cycle. Add 5 ml of a sterile solution of 2 M $MgCl_2$ just before use (19 gm of $MgCl_2$ in 90 ml of deionized H_2O and adjust to 100 ml). The prepared solution is SOB medium.

Sol. B :	Glucose	= 18 gm/litre
	Deionized H_2O	= 90 ml/litre

Mix well to dissolve the glucose and make the volume of solution to 20 ml. Filter solution through 0.22 µm filter and sterilize.

After the sterilization of Sol. – A (SOB) and it cool to 40°C or less add 20 ml of sterile 1 M solution of glucose (B).

AGAR MEDIUM/AGAROSE MEDIUM

First prepare liquid medium and add one of the following before sterilization.

Agarose/Bacto agar (for plate) = 15 gm/litre

Agarose/Bacto agar (for top agar) = 7 gm/litre

Sterilize by autoclaving for 20 minutes at 15 psi (1.05 kg/cm^2) or liquid cycle. After removal from the autoclave, swirl gently for uniform distribution of agar/agarose in liquid. Allow the medium to cool before addition of antibiotics or thermolabile chemicals. Plate can be poured directly from the flask. When the medium has hardened completely, invert the plates and store them at 4°C until needed. The plates should be removed from storage 1–2 hours before they are used.

LIQUID CULTURE (STORAGE MEDIA)

Bacteria growing on plates or in liquid culture, can be stored in aliquots of LB medium containing 30% sterile glycerol. Aliquots of 1 ml of LB with glycerol should prepared and vortexed to ensure that the glycerol is completely disperesd. Altimatively bacterial strains may be storec in LB freezing buffer.

LB freezing buffer :

- 36 mM K_2HPO_4 (anhydrous)
- 13.2 mM KH_2PO_4
- 1.7 mM Sodium citrate
- 0.4 mM $MgSO_4 \cdot 7H_2O$
- 6.8 mM ammonium sulfate
- 4.4% (v/v) glycerol in LB.

LB freezing buffer is made by dissolving the salts in 100 ml of LB to the specific concentrations. Measure 95.6 ml of the resulting solution into a fresh container, and then add 4.4 ml of glycerol. Mix the solution well and then sterilize by passing it through a 0.45 μm disposable N algene filter.

STAB CULTURE (STORAGE MEDIA)

Prepare stab culture in glass vials (2–3 ml) with screw or caps fitted with rubber glaskets. Add molten LB agar until the vials are two thirds full. Autoclave the partially filled vials for 20 minutes at 15 psi or liquid cycle. Remove the vial from autoclave, let them cool to room temperature, and then tighten the caps. Store the vials at room temperature until needed.

YPD (YEPD) MEDIUM

YPD is a complex medium for routine growth of yeast

- Yeast Extract = 10 gm
- Peptone = 20 gm
- Glucose = 20 gm
- H_2O = to 1000 ml

To prepare plates, add 20 gm of Bacto Agar (2%) before autoclaving.

TRIS HCl BUFFER

pH	Volume of 0.1 N HCl (ml)
1.1	45.7
1.2	44.7
1.3	43.4
1.4	42.0
1.5	40.3
1.6	38.5
1.7	36.6
1.8	34.5
1.9	32.0
1.10	29.2

GOOD BUFFER (NORMAN GOOD)

ACRONYM	CHEMICAL NAME (pH)
HEPES	N- (2-hydroxyethyl) piprazine-N-(2-ethane sulfonic acid) pH 6.8-8.2
Tricine	N-tris (hydroxymethyl) methylglycin (7.4-6.8)
BES	N-Nbis (2-hydroxyethyl)-2-aminoethanesulfonic acid (pH = 6.4-7.8)
Bis-Tris	bis (2-hydroxyethyl) imino tris (hydroxymethyl) methane (pH = 5.8-7.2)
TEA	Triethanolamine (pH = 7.3-8.3)
TES	N tris (hydroxymethyl) methyl-2-aminoethanesulfonic acid (pH = 6.8-8.2)
MES	2 (N-morpholino) ethanesulfonic acid (pH = 5.5-6.7)
Bis Tris propane	1,3-bis [Tris (hydroxymethyl) methyl amino] propane (pH = 6.3-9.5)

Phosphate Buffer's (Gomori Buffers)

Preparation of 0.1 M Potassium Phosphate Buffer

pH	Vol.1M K_2HPO_4(ml)	Vol.1M KH_2PO_4(ml)
6.0	13.2	86.8
6.2	19.2	80.8
6.4	27.8	72.2
6.6	38.1	61.9
6.8	49.7	50.3
7.0	61.5	38.5
7.2	71.7	28.3
7.4	80.2	19.8
7.6	86.6	13.4
7.8	90.8	9.2

Preparation of 0.1 M Sodium Phosphate Buffer

pH	Vol.1M Na_2HPO_4(ml)	Vol.1M NaH_2PO_4(ml)
6.0	12.0	88.0
6.2	17.8	82.2
6.4	25.5	74.5
6.6	35.2	64.8
6.8	46.3	53.7
7.0	57.7	42.3
7.2	68.4	31.6
7.4	77.4	22.6
7.6	84.5	15.5
7.8	89.6	10.4

Phosphate Buffered Saline (PBS)

- Distilled H_2O = 800 ml
- NaCl = 8 gm
- KCl = 0.2 gm
- Na_2HPO_4 = 1.44 gm
- KH_2PO_4 = 0.24 gm

Mix well and adjust pH to 7.4 with HCl. Add H_2O to 1 lit. Dispense the solution into aliquots and sterilize. Store the buffer at room temperature.

10 X Tris EDTA (TE)

- 100 mM Tris Cl
- 10 mM EDTA

pH of buffer is determined by pH of Tris Cl. Sterilize solutions and store at room temperature.

Tris–Cl (1M)

Dissolve 121.1 gm of Tris base in 800 ml of H_2O. Adjust the pH to the desired value by adding conc. HCl.

pH	HCl
7.4	70 ml
8.6	60 ml
8.0	42 ml

Adjust the volume of the solution to 1 lit with H_2O. If yellow color is appeared, use better quality of Tris.

Tris Buffered Saline (TBS)

- Distilled H_2O = 800 ml
- NaCl = 8 gm
- KCl = 0.2 gm
- Tris base = 3 gm

Mix well and add 0.015 gm of phenol red and adjust the pH = 7.4 with HCl. Rise volume upto 1 lit.

Lysozyme

Dissolve solid lysozyme at a concentration of 10 mg/ml in 10 mM Tris Cl (pH = 8.0) immediately before use. Make sure that pH is 8.

Pancreatic DNase I (1 mg/ml)

Dissolve 2 mg of crude pancreatic DNase I in 1 ml of (stock)

- 10 mM Tris–Cl (pH = 7.5)
- 150 mm NaCl
- mM $MgCl_2$

Dilution Solution

- 10 mM Tris–Cl (pH = 7.5)
- 150 mM NaCl
- 1 mM $MgCl_2$

When the DNase I is dissolved, add 1 ml of glycerol to the solution and mix by gently inverting the closed tube several times. Take care to avoid creating bubles and foam. Store the solution in aliquots of –20°C.

Pancreatic RNase (1 mg/ml)

Dissolve 2 mg of crude pancreatic RNase in 2 ml of TE.

Proteinase K (20 mg/ml)

Dissolve lyophilized powder at a concentration of 20 mg/ml in sterile 50 mM Tris (pH = 8.0), 1.5 mM calcium acetate. Divide the stock solution into small aliquots and store at –20°C. Each aliquots can be thawed and refrozen several times but should then be discarded. Unlike much cruder preparations of protease (*e.g.*, Pronase), proteinase K need not be self-digested before use.

Trypsin

Prepare bovine trypsin at a concentration of 250 µg/ml in 200 mM ammonium bicarbonate (pH = 8.9). Store the solution in aliquots at – 20°C.

10 X Amplification Buffer

- 500 mM KCl
- 100 mM Tris–Cl (pH = 8.3)
- 15 mM $MgCl_2$

Autoclave the 10X buffer for 10 minutes at 15 psi (1.05 kg/cm^2) or liquid cycle. Divide the sterile buffer into aliquots and store them at –20°C.

5 X BAL 31 Buffer

- 3 M NaCl
- 60 mM $CaCl_2$
- 60 mM $MgCl_2$
- 100 mM Tris–Cl (pH = 8.0)
- 1 mM EDTA (pH 8.0)

10 X Dephosphorylation Buffer (for use with SAP)

- 200 mM Tris–Cl (pH = 8.8)
- 100 mM $MgCl_2$
- 10 mM $ZgCl_2$

1 X Eco RI Methylase Buffer

- 50 mM NaCl
- 50 mM Tris-Cl (pH = 8.0)
- 10 mM EDTA
- 80 μM S-adenosylmethionine
- Store the buffer in small aliquots at –20°C.

10 X Exonuclease III Buffer

- 660 mM Tris–Cl (pH = 8.0)
- 66 mM $MgCl_2$
- 100 mM β-mercaptoethanol
- Add β-mercaptoethanol just before use

10 X proteinase K Buffer

- 100 mM Tris-Cl (pH = 8.0)
- 50 mM EDTA (pH 8.0)
- 500 mM $NaCl_2$

RNase H Buffer

- 20 mM Tris-Cl (pH = 7.6)
- 20 mM KCl
- 0.1 mM EDTA (pH 8.0)
- 0.1 mM dithiothreitol

Prepare fresh just before use

10 X Universal KGB (Restriction Endonuclease) Buffer

- 1M potassium acetate
- 250 mM Tris-acetate (pH 7.6)
- 100 mM magnesium acetate tetrahydrate
- 5 mM β-mercaptoethanol
- 0.1 mg/ml bovine serum albumin

Store the 10 X buffer in aliquots at –20°C

Alkaline Transfer Buffer (for alkaline transfer of DNA to Nylon membranes)

- 0.4 N NaOH
- 1 M NaCl

Neutralization Buffer II (for alkaline transfer of DNA to Nylon membrane)

- 0.5 M Tris–Cl (pH = 7.2)
- 1 M NaCl

Prehybridization Solution (for Dot, Slot and Northern Hybridization)

- 0.5 M Sodium phosphate (pH = 7.2)
- 7% (w/v) SDS
- 1 mM EDTA (pH = 7.0)
- 0.5 M phosphate buffer is 134 g of $Na_2HPO_47H_2O$, 4 ml of 85% H_3PO_4, H_2O to 1liter.

Acrylamide Solution (45%)

- Acrylamide (DNA sequencing gel) - 434 gm
- N,N'-methylenebiacrylamide - 16 gm
- Water - 600 ml

Heat the solution to 37°C to dissolve the chemicals. Adjust the volume to 1 liter with distilled water. Filter the solution through a nitrocellulose filter (0.45 μ pore size) and store filtrate in dark bottle at room temperature.

Ethidium bromide (10 mg/ml)

Add 1 gm of ethidium bromide to 100 ml of distilled water, stir on a magnetic stirrer for several hours to ensure that the dye has dissolved. Wrap the container in aluminum foil or transfer the solution to a dark bottle and store at room temperature.

Phenol

Commonly the commercial liquefied phenol is clear and colorless. It can be used in molecular cloning without redistillation. Occasionally a liquefied phenol is pink or yellow. It is not fit for experimental works because it contains oxidation products. The oxidation products are removed by redistillation at 160°C.

The equilibrium of phenol in 7.8 pH is also an important step because the DNA is partitioned into the organic phase at acidic pH. It requires storage of liquefied phenol at -20°C. At the time of need phenol is removed from the freezer and warm at room temperature and then melt it at 68°C. Add hydroxyquinoline to a final concentration of 0.1%. The compound acts as antioxidant, a partial inhibitor of RNase and a week chelater of metal ion.

To the melted phenol add an equal volume of buffer (usually 0.5 M Tris–Cl, pH = 8.0) at room temperature and stir the mixture for 15 minutes. The phenol is separated into two phases. The upper aqueous phase is aspirated and the extraction process is repeated until the pH of phenolic phase is not reached at pH > 7.8 (measured with pH paper). The 0.1 volume of 0.1 M Tris–Cl containing 0.2% mercaptoethanol is added in aqueous phase. The solution is kept in light tight bottle at 4°C for one month.

IPTG (20% w/v, 0.8 M)

Dissolve 2 gm of isopropylthio – β – D-galactoside in 8 ml of distilled water. Adjust the volume of solution to 10 ml with water and sterilize by passing it through a 0.22 μm disposable filter.

APPENDIX–2

SOLUTION PREPARATION

A solution is the system in which a substance is evenly distributed throughout a liquid without any tendency towards separation. The dispersing liquid is called solvent and the dispersed substance is called the solute. Usually a solute is a solid substance but it may be liquid or a gas. A solution is characterized by its homogenoneity, no tendency of separation, passage through a filter, single-phase system, no boundary zone, indistinguishable even with ultra microscope with respect to both components of solution.

Commonly the solution prepared by mixing of weak acid and its sodium or potassium salt is known as buffer. The buffer solution maintains its actual pH even after mixing of considerable volume of acids and alkalies at a specific temperature.

In the experiments the solution or reagent is sometimes represented by its strength. The strength of solution is determined by the amount of solute present in solution. The commonly used solution strengths are as follows:

1. Percent Solution
2. Part dilution
3. Molar solutions
4. Normal solutions.

PERCENT SOLUTIONS

It is most common way to express solution strength. Percent represents per hundred units solvent, the weight solute in part of final solution. Percent solution can further be express by three processes.

(*a*) Weight per unit weight (w/w)

(*b*) Weight per unit volume (w/v)

(*c*) Volume per unit volume (v/v)

(a) Weight per unit weight (w/w)

It represents that the weight of both solute and solvent add up to 100 without considering the final volume of solution.

(b) Weight per unit volume (w/v)

It is a common method for preparation of laboratory solutions. For preparation of solution, a considerable weight of the solute is dissolved in a final volume of 100 ml. Some persons commit mistake by dissolving a considerable weight of solute in 100 ml of solvent. A volumetric flask is the most suitable and convenient container for preparing the solution. Put the weighted solute in the flask and add the solvent (always keep less volume than required final volume). After the complete mixing of solute add more solvent to get desired volume of solution.

(c) Volume per unit volume (v/v)

It is used when the solution is prepared by the mixing of two liquids. A known volume of the liquid solute is added to liquid solvent to make final volume of 100 ml. In the preparation of solution volumetric flask is used.

PART DILUTION

In this method the dilution is expressed as that part of the whole volume into which one part is dissolved. For example, WBC dilution 1:20 represents that in total volume one part is blood in per 20 part of solution. Hence the actual volume of diluting solution is 19 parts per one part of blood.

MOLAR SOLUTION

A molar solution contains the molecular weight of the solute in gram per liter of solution. The molecular weight is calculated by addition of the atomic weight of the different atoms present in compound. For example, the molecular weight of NaCl in 58.454 gm. Hence 1 molar solution of the salt contains 58.454 gm of NaCl in 1000 ml (1 M = 1000 mM).

$$\text{Molarity} = \frac{\text{Number of moles of solutes}}{\text{Volume of solution in liter}}$$

Formula to convert a percent solution into a mol/L solution is mentioned below.

$$\text{Mol/L solution} = \frac{\text{gm\% (w/v) sol} \times 10}{\text{mole weight of solute}}$$

Formula for the conversion of a normal solution into mol/L solution is mentioned below.

$$\text{Molarity} = \frac{\text{Normality of solution}}{\text{Valence of substance}}$$

Sometimes we want to mix two solutions of different molar strength or prepare a fresh solution from stock has different molar strength. When we mix two solutions have different molar strength, the molar strength of the resultant solution can be determined by formula

$$M_3 = \frac{M_1V_1+M_2V_2}{V_1+V_2}$$

In second condition the volume of water to be added in stock solution to get a working solution of known molar strength is calculated by formula

$$V_1 - V_2 = \left[\frac{M_1 - M_2}{M_2}\right] V_1$$

where

$V_1 - V_2$ = volume of water added

M_1 = molar strength of stock solution

M_2 = molar strength of working solution

V_1 = volume of stock solution

Illustration-A2–1

Calculate the weight of solutes requires for preparation of solution strength mentioned below, in 100 ml of solvent.

(a) 10 mM Tris (121.1 gm)

(b) 1 mM EDTA (375 gm)

(c) 100 Glucose (96 gm)

(d) 0.5 M EDTA (375 gm)

(*a*) 1M solution requires 121.1 gm in 1000 ml of water.

1 mM solution requires 121.1/1000 = 0.1211 gm in 1000 ml of water.

10 mM solution requires 0.1211 × 10 = 1.211 gm in 1000 ml of water.

Hence

1000 ml water requires 0.1211 gm

100 ml water requires 1.211 × 100/1000 = 0.1211 gm.

(*b*) 1M solution requires 375 gm in 1000 ml of water.

1 mM solution requires 375/1000 = 0.375 gm in 1000 ml of water.

Hence

1000 ml water requires 0.375 gm

100 ml water requires 0.375 × 100/1000 = 0.0375 gm

(*c*) 1M solution requires 96 gm in 1000 ml of water.

1 mM solution requires 96/1000 = 0.096 gm in 1000 ml of water.

100 mM solution requires 0.096 x 100 = 9.6 gm in 1000 ml of water.

Hence

1000 ml water requires 9.6 gm

100 ml water requires 9.6 × 100/1000 = 0.96 gm.

(*d*) 1M solution requires 375 gm in 1000 ml of water.

0.5M solution requires 375/2 = 187.5 gm in 1000 ml of water.

Hence

1000 ml water requires 187.5 gm

100 ml water requires 187.5 × 100/1000 = 18.75 gm.

Illustration-A2-2

In a laboratory 10 mM stock solution of EDTA is available. A person gets 20 ml and wants to prepare 2 mM working solution. Calculate the volume of water necessary for preparation of working solution.

$$V_1 - V_2 = \left[\frac{M_1 - M_2}{M_2}\right] V_1$$

$$V_1 - V_2 = \left[\frac{10-2}{2}\right] 20 = 80 \text{ ml}$$

Illustration-A2-3

The 150 ml of 10 mm glucose solution and 15 ml of 2 mm glucose solutions are mixed. Calculate the strength of resultant solution.

$$M_3 = \frac{M_1V_1 + M_2V_2}{V_1 + V_2}$$

$$M_3 = \frac{10 \times 150 + 2 \times 15}{150 + 15} = 9.3 \text{ Mm}$$

NORMAL SOLUTION

A normal solution contains the gram equivalent weight of a solute per liter of solvent. The equivalent weight is the number of units of the substance, which will combine with or replace a single unit of hydrogen, 35.5 unit of chloride, 8 unit of oxygen etc. The methods for calculation of equivalent weight are mentioned below.

(*a*) The equivalent weight of an element is calculated by dividing the atomic weight by valency. For example, equivalent weight of sulphur (atomic weight of S = 32.006 and valency = 2) is 16.003.

(*b*) The equivalent weight of an acid is the weight of an acid in gram, which contains 1.008 gm of replaceable hydrogen. It is calculated by dividing the molecular weight by the number of replaceable hydrogen atoms in the molecule. For example, the molecular weight of H_2SO_4 (molecular weight = 98.082 and number of replaceable hydrogen = 2) is 49.041.

(*c*) The equivalent weight of an alkali is calculated the equivalent weight of an acid neutralized. It is calculated by dividing the molecular weight by the number of hydroxyl redicals in the molecules. For example, the equivalent weight of calcium hydroxide (molecular weight = 91.014 and the number of hydroxyl redical = 2) is 45.552.

(*d*) The equivalent of salt is calculated by dividing its molecular weight by the number of metal ions (cation) per molecules, multiplied by the valence of the cations. For example, the equivalent weight of Na_2SO_4 (molecular weight = 142.060) is

Valency of Na^+ = 1, Number of cations = 2

$$Na_2SO_4 = \frac{142.062 \times 1}{2} = 71.030$$

Appendix–3

References

1. Lilley, D.M.J. (1989): *Chem. Soc. Rev.*, 18, pp-53–83.
2. Aaij, C and Borst, P (1972): The Gel electriphoresis of DNA, *Biochem. Biophys. Acta.*, 269, pp-192 – 200
3. Backman, K (1980): A cautionary note on the use of certain restriction endonuclease with methylated substrates. *Gene.* 11, pp-169–171
4. Bajorath, J; Hinrichs, W and Saenger, W (1988): The enzymatic activity of proteinase K is controlled by calcium. *Eur. J. Biochem.* 176, pp-441–447
5. Bassam, B.J; Caetano-Aumolles, G and Gresshoff, P.M (1991): Fast and sensitive silver staining of DNA in polyacrylamide gels. *Anal. Biochem.*196, pp-80–83
6. Berch, M. J. (ed.) (1991): *The ACT Cytogenetics Laboratory Manual.* Second edition, Reven Press, New York.
7. Birren, B; Green, E.D.; Klapholz, S; Mayers, R.M and Roskams, J (1997): *Genome Analysis: A Laboratory Manual*, Vol-1, Analyzing DNA (ed.E.D.Green et al) Cold Spring Harbor Laboratory Press, New York
8. Bjerrum, O.J and Schafer-Nielsen, C(1986): Buffer systems and transfer parameters for semi dry electroblotting with a horizontal apparatus. In *Electrophoresis 86: Proceeding of the 5th meeting of the International Electrophoresis Society* (ed.M.J.Dunn), pp- 315–327, VCH Publishers, Deerfield Beach, Florida.
9. Blackburn P.,Wilson G., and Moore S. (1977): Ribonuclease inhibitor from human placenta. Purification and properties. J Biol Chem.252,pp-5904–5910.
10. Blake M.S., Johnston K.H., Russell-Jones G.J., and Gotschlich E.C. (1984): A rapid sensitive method for detections of alkaline phosphatase-conjugated anti-antibody on Western blots. Anal Biochem. 136, pp-175–179.
11. Bolivar F. and Backman K. (1979): Plasmids of Escherichia coli as cloning vectors. Methods Enzymol. 68: pp 245–267
12. Bonhoeffer S., Mc Caskill J.S., Stadter P.F., and Schuster P.(1993): RNA multistructure landscape . Biophys. J. 22: pp 13–24.

13. Bronstein I., Voyta J.C., Lazzari K.G., Murphy O., Edwards B., and Kricka L.J.(1990): Rapid and sensitive detection of DNA in Southern blots with chemiluminecence. *Bio Techniques 8: pp-310–314.*

14. Bronstein I., Voyta J.C.,Murphy O.J., Tizard R., Ehrenfels C.W., and Cate R.L. (1993).Detection of DNA in Southern blots with chemiluminescence. *Method Enzymol. 217. pp 398–414.*

15. Burge C.B. and Karlin S. (1997): Predection of complete gene structures in human genomic DNA. *J.Mol.Biol.268,pp 78–94.*

16. Chomezynski,P and Sacchi, N (1990): Single step RNA isolation from cultural cells and tissue. In "*Current Protocol in Molecular Biology*" (F.M.Ausbel et al ed.) Vol-1, pp-424 to 428, Greene. Publ. Assoc. and Wiley, Interscience, New York.

17. Chow, C. S. and Barton, J, K. (1990): *J. Am. Chem. Soc.*, 112, 2839–2841.

18. Chrambach A. and Rodbard D. (1972). Polymerization of Polyacrylamide gels- Efficiency and reproducibility as a function of catalyst concentration. *Biochemistry*, 31, pp-11665–11676.

19. Davidsons J.N. (1972): *The biochemistry of the nucleic acids*, 7th edition, Academic Press. New York.

20. Deininger P.L (1983): Approaches to rapid DNA sequence analysis. *Anal. Biochem*, 135, pp-247–263.

21. Eschenfeldt W.H., Pukas R.S., and Berger S.L. 1987. *Homopolymeric tailing. Methods Enzymology*, 52, pp-337 – 343.

22. Fiechter, A; Kappeli, O and Menssdoerffer, F (1987): *Batch and Continuous Culture. In The Yeast* (A.H.Rose, A,E.Wheals and J.S.Harrison ed.). Vol-2. Academic Press, San Diego, pp-99–130.

23. Gaillard C. and Strauss F (1990): Ethanol precipitation of DNA with linear gels, acrylamide as a carrier. *Nucleic Acid Res*, 18, pp–378.

24. Guschibauer, W; Chantot, J.F and Theile, D (1990): *J. Biomol. Struct. Dynamics*, 8, pp-491–511

25. Hall, R. H. (1971): *The Modified Nucleiosides In Nucleic Acids*, Columbia University Press, New York.

26. Hanahan, D (1983): *Studies on transformation of Escherichia coli with Plasmids*. J. Mol. Biol, 166, pp- 557–580.

27. Htun, H and Dahlberg, J.E (1988): *Science*, 241, pp-1791 – 1796

28. Johansson S. and Skoog B. (1987): Rapid silver staining of polyacrylamide gels. *J. Biochem and Biophy*. Methods (supp),14, pp-53

29. Kirby K.S. (1956):A new method for the isolation of ribonucleic acids from mammalian tissues. *J. Biochem*, 64, p-405

30. Kunitz M. (1950): Crystalline deoxyribonuclease II. Digestion of thymus nucleic acid (deoxyribonucleic acid). The kinetics of the reaction. *J. Gen. Physiol*, 33, pp-363–377.

31. Labarca C. and Paigen K. (1980): A simple, rapid and sensitive DNA assay procedure . *Anal. Biochem*, 02, pp-344 – 352.

32. Lehrach H., Diamond D., Wozney J.M., and Boedtker H. (1977): RNA molecular weight determinations by gel electrophoresis under denaturing conditions, a critical re-examination. *Biochemistry*, 16, pp-4743–4751.

33. Lindsay S. and Bird A.P. (1987): Use the restriction enzymes to detect potential gene sequences in mammalian DNA. *Nature*, 327, pp-337–338.

34. Lumn G. and Sansone E.B. (1987): Ethidium bromide; destruction and decontamination of solutions. *Anal. Biochem.* 162, pp-453–458.

35. Manchester K.L (1995): Value of $A_{260/280}$ ratios for measurement of purity of nucleic acids. *Bio Technique*, 19, pp-208–219

36. Manchester K.L (1996): Use of UV methods for the measurement of protein and nucleic acids. *Bio Technique*. 20, pp-268–270.

37. Matsudaria, P; and Burgess, D.R (1978): SDS microslab linear gradient polyacrylamide gel electrophoresis. *Anal. Biochem.* 87, pp-383–396.

38. Merril C.R.(1990): Silver staining of proteins and DNA. *Nature*, 343, pp-779–780.

39. Miller J.H (1972): *Experiments in molecular genetics*. Cold Spring Harbor Laboratory, Cold Spring Harbor, New York.

40. Mitchell L.G., Bodenteich A., and Merril C.R. 1994. Use of silver staining to detect nucleic acids. *Methods Mol Biol*.31, pp-197–203.

41. Oakely B.R., Kirch D.R., and Morris. N.R. 1980. A simplified ultrasensitive silver stain for detecting proteins in polyacrylamide gels. *Anal. Biochem* 105, pp-361–363.

42. Pollard-Knight D., Simmonds A.C., Schaap A.P., Akhavan H., and Brady M.A.(1990): Nonradioactive DNA detection on Southern blots by enzymatically triggered chemiluminescence. *Anal. Biochem*.185, pp-353–358.

43. Primrose, S. B; Twyman, R. M and Old, R. W. (2001): *Principles of Gene Manipulation*, Blackwell Publishing Company, MA 02148–5020, USA.

44. Quillardet P. and Hofnung M. (1988): Ethidium bromide and safety-Readers suggest alternative solution (letter). *Trends Genet.* 4, pp-89–90.

45. Rajgopal,P. and Feigon, J (1989): *Nature*, 339, pp-637–640.

46. Raleigh E.A. and Wilson G. 1986. *Escherichia coli* K-12 restricts DNA containing 5-methylcytosine. *Proc. Natl. Acad. Sci.* 83, pp-9070–9074.

47. Richterich P. and Church G.M. 1993. DNA sequencing with direct transfer electrophoresis and nonradioactive detection. *Methods Enzymol.* 218, pp-187–222.

48. Roomey, D, E and Czepulkowski, B. H (ed.) (1992): *Human Cytogenetics*, Second edition, Vol-1, Oxford University Press, New York.

49. Rould, M.A; Perona, J.J; Soll, D and Steitz, T.A (1989): *Science*, 246, pp-1135–1142.

50. Roychoudhary R. and Wu R. (1980). Terminal transferase catalyzed addition of nucleotides to the 3-termini of DNA. *Methods Enzymol.* 65, pp-43–62.

51. Ruther U. and Muller-Hill B. (1983). Easy identification of C-DNA clones *EMBO J.*, 2, pp-1791 – 1794.

52. Saenger, W (1984): *Principle of Nucleic Acid Structure*, Springer Verleg, New York.

53. Sambrook, J; Fritsch, E.F and Maniatis, T (1989): Extraction and Purification of RNA. In "*Molecular Cloning*" Vol-1,Cold Spring Harbor Laboratory Press, New York.

54. Schimmel, P. (1989): *Cell*, 58, pp-9–12.

55. Semenza G. (1957). Chromatography of deoxyribonucleic acid in calcium phosphate. *Ark. Kemi*.11, pp-89–96.

56. Smith H.O (1980). Recovery of DNA from gels. *Methods Enzymol*,65, pp-371–380.

57. Southern E.M. 1975. Detection of specific sequence among DNA fragments separated by gel electrophoresis. *J. Mol. Biol*. 98, pp-503–517.

58. Spector D.L., Goldman R.D., and Leinwad L.A. (1998). *Cells: A laboratory manual* vol 1., culture and biochemical analysis of cells.. pp 2.8–2.10. Cold Spring Harbor Laboratory Press, Cold Spring Harbor (New York.)

59. Sutcliffe J.G. (1979). Complete nucleotide sequence of the Escherichia coli plasmid pBR-322. *Cold Spring Harbor Symp. Quant. Biol*.43, pp-77–90.

60. Svaren J., Inagami S., Lovegren E., Chalkley. (1987). DNA denatures upon drying after ethanol precipitation. *Nucleic Acids Res*.15, pp-8739–8754.

61. Tartof K.D. and Hobbs. C.A. 1987. Improved media for growing plasmid and cosmid clones. *Focus (Life Technology)*.9, p-12.

62. Tung W.L. and Chow K.C. 1995. A modified medium for efficient electrotransformation of E.coli. *Trends Genet*. 01, pp-128–129.

63. Van F. and Bell K. 1996. A simplified silver diamine method for the staining of nucleic acids in polyacrylamide gels.*Electrophoresis*, 17, pp-20 – 25.

64. Van Oss. C.J., Good R.J., and Chaudhary M.K. 1987. Mechanisms of DNA (Southern) and proteins (Western) blotting on cellulose nitrate and other membrances. *J. Chromatogr*. 319, pp-53–65.

65. Verma, R.S and Babu, A (1989): *Human Chromosomes*, Pergamon Press, New York.

66. Waring M.J. 1965. Complex formation between ethidium bromide and nucleic acids. *J. Mol. Biol*.13, pp-269–282.

67. Waring M.J. 1966.. Structural requirements for the binding of ethidium bromide to nucleic acids. *Biochem. Biophys. Acta* 114.pp-234–244.

68. Waring, M. J. (1970): J. Mol. Biol., 54, pp-247–279.

69. Wilfinger W.W. Mackey K., and Chomczynski P. 1997 . Effect of pH and ionic strength on the spectrophotometric assessment of nucleic acid purity. *Biotechniques*, 22, 474–488

70. Yanofsky, C. (1988): *J. Biol. Chem*., 263, pp-609–612.

INDEX